T0297931

FROM PHOTONS TO HIGGS
A Story of Light (2nd Edition)

FROM PHOTONS TO HIGGS
A Story of Light (2nd Edition)

Moo-Young Han
Korea Advanced Institute of Science and Technology (KAIST), Korea
& Duke University, USA

 **World Scientific**

NEW JERSEY · LONDON · SINGAPORE · BEIJING · SHANGHAI · HONG KONG · TAIPEI · CHENNAI

Published by

World Scientific Publishing Co. Pte. Ltd.

5 Toh Tuck Link, Singapore 596224

USA office: 27 Warren Street, Suite 401-402, Hackensack, NJ 07601

UK office: 57 Shelton Street, Covent Garden, London WC2H 9HE

Library of Congress Cataloging-in-Publication Data
Han, M. Y., author.
  [Story of light]
  From photons to Higgs : a story of light / Moo-Young Han, Korea Advanced Institute of Science
and Technology (KAIST), Korea, Duke University, USA. -- 2nd edition.
    pages cm
  Revision of: A story of light. 2004.
  ISBN 978-9814579957 (hardcover : alk. paper) -- ISBN 9789814583862 (softcover : alk. paper)
  1. Particles (Nuclear physics) 2. Quantum field theory. I. Title.
  QC793.2.H36 2014
  530.14'3--dc23
                          2013045018

British Library Cataloguing-in-Publication Data
A catalogue record for this book is available from the British Library.

Credit for Cover Image:
Photograph produced by CERN: "CMS: Simulated Higgs to two jets and two electrons".
© 1997 CERN (License: CC-BY-SA-3.0)
Image creator: Lucas Taylor

Typeset by Stallion Press
Email: enquiries@stallionpress.com

Printed in Singapore

For Eema, Grace, Chris and Tony

Acknowledgments

I would like to express my gratitude to Dr. K. K. Phua, the Chairman of the World Scientific Publishing Company, for visionary encouragement and steady guidance throughout the work on this Second Edition.

I also appreciate the kindness and encouragement of Professor Sung Mo (Steve) Kang, the President of Korea Advanced Institute of Science and Technology (KAIST), who provided opportunity for the completion of this work while at KAIST.

Lastly, thanks are also due to Mr. Kah Fee Ng for his meticulous editing of the manuscript that made significant improvement on this book.

Foreword to the Second Edition

When the first version of this book was published in 2004, the Tevatron collider near Chicago, then the world's highest-energy particle accelerator, has peaked to its maximum energy. It helped to pin down the top quark which completed the picture of three generations of quarks and leptons. Its highest energy was however not sufficient to find evidences for the long sought-after Higgs particle.

Without either the confirmation or refutation of the Higgs mechanism, the Standard Model of quarks and leptons for the electroweak interaction was more or less on hold, waiting for more powerful particle accelerator to come online. The particle physics community was eagerly waiting for the successful completion of the Large Hadron Collider (LHC), which was under construction at CERN, the Center for European Nuclear Research, in Geneva, Switzerland.

Finally, after a long wait, the LHC came online in 2009 and on July 4, 2011, the dramatic announcement was made to the effect that researchers at CERN have accumulated enough evidence for a particle that looked very much like the long sought-after Higgs particle.

Since then more supporting data were accumulated in favor of the newly discovered particle being the Higgs particle. But to this day people are still hedging. They refer to it as a "Higgs-like particle" or "a" Higgs particle but not "the" Higgs particle and so on. All-out efforts to pin down the Higgs particle are now one of the most — if not the most — urgent tasks of the particle physics.

Higgs particles are the quanta of the Higgs field which is responsible for the spontaneously broken symmetry critically needed for the Standard Model for the electroweak force. The concept and the mechanism of the spontaneously broken symmetry within the framework of the gauge field theory are, however, rather difficult to grasp and hard to explain to those outside the immediate circle of specialists. At one time, a cabinet minister in the United Kingdom even offered a hefty prize to anyone who can explain the ideas behind the Higgs particle to nonspecialists.

In this updated and expanded version, I will attempt to tackle this task. I will strive to explain in simple terms 1) What are the nuts and bolts of the spontaneously broken gauge symmetry, 2) What exactly is the Higgs particle, and 3) How it helps to provide a mathematical technique that can be viewed as giving mass to particles that were massless to start with.

The Standard Model represents the extension of the local gauge symmetry inherent in electromagnetic interaction to the cases of weak and strong nuclear forces. As such, it corresponds to the third time we have emulated the properties of the electromagnetic interaction and the photons.

The development of quantum mechanics was in essence adaptation of the wave-particle duality of photons to all matter particles with mass; likewise, the formulation of the quantum field theory was emulation of the successful theory of the quantization of the electromagnetic field.

Viewed this way, all the new developments in the 20th and 21st century — quantum mechanics, quantum field theory, gauge field theory, spontaneously broken symmetry, and the Higgs mechanism — correspond to continuing the story of emulation of the electromagnetic interaction and photons, to wit, "A Story of Light."

In this second edition, eight new chapters have been added, Chapters 12 through 19, to replace the original three chapters, Chapter 12, 13, and 14. The first eleven chapters that deal with the basics of the Lagrangian quantum field theory, as well as the original Prologue, remain the same as in the first edition. The new chapters address current topics in local gauge field theories including the electroweak theory, spontaneous symmetry breaking and the Higgs mechanism, history of the color charges of quarks and the quantum chromodynamics.

Contents

Prologue

The relativistic quantum field theory, or quantum field theory (QFT) for short, is the theoretical edifice of the standard model of elementary particle physics. One might go so far as to say that the standard model *is* the quantum field theory. Having said that as the opening statement of this book, we must be mindful that both quantum field theory and the standard model of elementary particle physics are topics that are not necessarily familiar to many individuals. They are subject areas that are certainly not familiar to those outside the specialty of elementary particle physics, and in some cases not too well grasped even by those in the specialty.

The Standard Model of elementary particle physics is a term that has come into prominence as it became the paradigm of particle physics for the last three decades. In brief, the standard model aims to understand and explain three of the four fundamental forces — the electromagnetic, strong nuclear and weak nuclear — that define the dynamics of the basic constituents of all known matter in the universe.[1] As such, it consists of two interrelated parts: the part

---

[1]The fourth force of nature, gravity, does not come into play in the scale of the mass of elementary particles and is not included in the standard model. Attempts

that deals with the question of what are the basic building blocks of matter and the second part concerned with the question of what is the theoretical framework for describing the interactions among these fundamental constituents of matter.

A century after the original discovery of quantum of light by Max Planck in 1900 and its subsequent metamorphosis into photon, the zero-mass particle of light, by Albert Einstein in 1905, we have come to identify the basic constituents of matter to be quarks and leptons — the up, down, strange, charm, top and bottom, for quarks, and the electron, muon, tauon, electron-type neutrino, muon-type neutrino, and tauon-type neutrino, for leptons. The three forces are understood as the exchange of "quanta" of each force — photons for the electromagnetic force, weak bosons for the weak nuclear force, and gluons for the strong nuclear force. These particles, some old, such as photons and electrons and some relatively new, such as the top and bottom quarks or the tauons and their associated neutrinos, represent our latest understanding of what are the basic constituents of known matter in the universe.

There are scores of books available which discuss the basic particles of matter, at every level of expertise. For a general readership, we can mention two books that contain no or very little mathematics, *Quarks and Gluons* by myself and *Facts and Mysteries in Elementary Particle Physics* by Martinus Veltman.[2]

The theoretical framework for the three forces or interactions is quantum field theory, that is, the relativistic quantum field theory. Each force has its own form, and again, some old and some new. Quantum electrodynamics, QED for short, was fully developed by the end of the 1940s and is the oldest — and more significantly, the only truly successful quantum field theory to date — of the family. Quantum chromodynamics, QCD, is the framework for the strong nuclear force that is mediated by exchanges of gluons. It was initiated

---

to merge gravity with the standard model have spawned such ideas as the grand unified theory, supersymmetry, and supersting, the so-called theory of everything. These topics are not discussed in this book.

[2] *Quarks and Gluons* by M. Y. Han, World Scientific (1999); *Facts and Mysteries in Elementary Particle Physics* by Martinus Veltman, World Scientific (2003).

in the 1960s and has been continually developed since, but it is far from becoming a completely successful quantum field theory yet. The theory for the weak nuclear force, in its modern form, was also started in the 1960s, and in the 1970s and 1980s, it was merged with quantum electrodynamics to form a unified quantum field theory in which the two forces — the electromagnetic and weak nuclear — were "unified" into a single force referred to as the electroweak force. Often this new unified theory is referred to as the quantum flavor dynamics, QFD. Thus, the quantum field theory of the standard model consists of two independent components — quantum chromodynamics and quantum flavor dynamics, the latter subsuming quantum electrodynamics.

Despite the abundant availability of books, at all levels, on basic building blocks of matter, when it comes to the subject of relativistic quantum field theory, while there are several excellent textbooks at the graduate level, few resources are available at an undergraduate level. The reason for this paucity is not difficult to understand. The subject of quantum field theory is a rather difficult one even for graduate students in physics. Unless a graduate student is interested in specializing into elementary particle physics, in fact, most graduate students are not required to take a course in quantum field theory. It is definitely a highly specialized course. Quantum field theory thus remains, while a familiar term, a distant topic. Many have not had the opportunity to grasp what the subject is all about, and for those with some rudimentary knowledge of physics at an undergraduate level beyond the general physics, the subject lies well beyond their reach.

The main purpose of this book is to try to fill this gap by bringing out the conceptual understanding of the relativistic quantum field theory, with minimum of mathematical complexities. This book is not at all intended to be a graduate level textbook, but represents my attempt to discuss the essential aspects of quantum field theory requiring only some rudimentary knowledge of the Lagrangian and Hamiltonian formulation of Newtonian mechanics, special theory of relativity and quantum mechanics.

There is another theme in this book and it is this. Throughout the course of development of quantum field theory, from the original quantum electrodynamics in which the Planck–Einstein photon is deemed as the natural consequence of field quantization to the present-day development of the gauge field theory for quarks and leptons, the theories of electromagnetic field have been — and continue to be — a consistently useful model for other forces to emulate. In this process of emulating theories of electromagnetic field, the concept of particles and fields would go through three distinct phases of evolution: separate and distinct concepts in classical physics, the particle-wave duality in quantum mechanics, and finally, particles as the quanta of quantized field in quantum field theory. As we elaborate on this three-stage evolution, we will see that the photon has been — and continues to be — the guiding light for the entire field of relativistic quantum field theory, the theoretical edifice of the standard model of elementary particle physics.

1

Particles and Fields I: Dichotomy

One may have wondered when first learning Newtonian mechanics, also called the classical mechanics, why the concept of a field, the force field of gravity in this case, is hardly mentioned. One usually starts out with the description of motion under constant acceleration — the downward pull of gravity with the value of $9.81\,\mathrm{m/s^2}$. Even when the universal law of gravity is discussed, for example, to explain the Kepler's laws, we do not really get into any detailed analyses of the force field of gravity.

In classical mechanics the primary definition of matter is the point mass, and the emphasis is on the laws of motion for point masses under the influence of force. The focus is on the laws of motion rather than the nature of force field, which is not really surprising when we consider the simplicity of the terrestrial gravitational force field — uniform and in one parallel direction, straight down toward the ground. A point mass is an abstraction of matter that carries mass and occupies one position at one moment of time and this notion of a point mass is diagonally opposite from the notion of a field, which, by definition, is an extended concept, spread out over a region of space.

As we proceed from the study of classical mechanics to that of classical electromagnetism, we immediately notice a big change; from day one it is all about fields. First the electric field, then the magnetic field, and then the single combined entity, the electromagnetic field. No sooner than the Coulomb's law is written down, one defines the electric field and its spatial dependence is determined by Gauss' Law. Likewise, Ampere's Law determines the magnetic field and finally the laws of Faraday and Maxwell lead to the spatial as well as temporal dependence of electromagnetic field.

This dichotomy of the concept of point particle and that of field is in fact as old as the history of physics. From the very beginning, back in the 17th century, there were two distinct views of the physical nature of light. Newton advocated the particle picture — the corpuscular theory of light — whereas Christian Huygens advanced the wave theory of light. For some time — for almost a century and half — these two opposing views remained compatible with what was then known about light — refraction, reflection, lenses, etc. Only when in 1801 Thomas Young demonstrated the wave nature of light by the classic double-slit interference experiment, with alternating constructive and destructive interference patterns, the wave theory triumphed over the particle theory of light.

One might have wondered why the notion of field did not play a prominent role in the initial formulation of Newtonian mechanics, especially since both the gravitational force law and the Coulomb's law obey the identical inverse square force law:

$$F = G\frac{m_1 m_2}{r^2} \quad \text{for gravity}$$

and

$$F = k\frac{q_1 q_2}{r^2} \quad \text{for Coulomb's law}$$

where G and k are the respective force constants, m is mass and q is the electric charge.

The disparity is simply a practical matter of scale. At the terrestrial level, in our everyday world, the inverse square law really does not come into play; the curvature of the surface of the earth is approximated by a flat ground and the gravitational force lines directed toward the center of the earth become, in this approximation, parallel lines pointing downward. In this scale of things, the field aspect of gravity is just too simple to be taken into account. There is no need to bring in any analyses of the gravitational field in the flat surface approximation.

On the contrary, with electric and magnetic forces, we notice and measure in the scale of tabletop experiments the spatial and temporal variations of these fields. The gradients, divergences and curls, to use the language of differential vector calculus, of the electric and magnetic fields come into play in the scale of the human-sized world and this is why the study of electromagnetism always starts off with the definition of electric and magnetic fields.

This well-defined dichotomy of particles and fields, diagonally opposite concepts in classical physics, would evolve through many twists and turns in the twentieth century physics of relativity and quantum mechanics, ending up eventually with the primacy of the concept of field over that of particle in the framework of quantum field theory.

The process of evolution of the concepts of particles and fields has taken a quite disparate path. The Newtonian mechanics has evolved through several steps, some quite drastic. First, there was the Lagrangian and Hamiltonian formulation of mechanics. One of the most important outcomes of this formalism is the definition of what is called the canonically conjugate momentum and this would pave the way for the transition from classical mechanics to quantum mechanics. Quantum field theory could not have developed had it not been the idea of canonically conjugate momentum defined within the Lagrangian and Hamiltonian formalism. As quantum mechanics is merged with special theory of relativity, the culmination of the particle view was reached in the form of relativistic quantum mechanical wave equations, such as the Klein–Gordon and Dirac equations, wherein the wavefunction solutions of these equations provide the relativistic quantum mechanical description of a particle. (More on these equations in later chapters.)

In contradistinction to this development of particle theory, the field view of classical electromagnetism remained almost totally unmodified. The equation of motion for charged particles in an electromagnetic field is naturally accommodated in the Lagrangian and

Hamiltonian formalism. In the Lagrangian formulation of classical mechanics, Maxwell's equations find a natural place by being one of the few examples of what is called the velocity-dependent potentials (more on this in the next chapter). The very definition of the canonically conjugate momentum for charged particles to be the sum of mechanical momentum and the vector potential of the electromagnetic field, discovered back in the 19th century, is in fact the foundation for quantum electrodynamics of the 20th century.

The contrast between the mechanics of particles and the field theory of electromagnetic fields becomes sharper when dealing with the special theory of relativity. The errors of Newtonian mechanics at speeds approaching the speed of light are quite dramatic, and of course, the very foundation of mechanics had to be drastically modified by the relativity of Einstein. Maxwell's equations for the electromagnetic field, on the other hand, required no modifications whatsoever at high speeds; the equations are valid for all ranges of speeds involved, from zero to all the way up to the speed of light. At first, this may strike as quite surprising, but the fact of the matter is that Maxwell's equations lead directly to the wave equations for propagating electromagnetic radiation — light itself. Maxwell's theory of the electromagnetic field is already fully relativistic and hence need no modifications at all.

The development of relativistic quantum mechanics demonstrates quite dramatically the primacy of the classical field concept over that of particles. To cite an important example, in relativistic quantum mechanics, the first and foremost wave equation obeyed by particles of any spin, both fermions of half-integer spin and bosons of integer spin, is the Klein–Gordon equation. Fermions must also satisfy the Dirac equation in addition to the Klein–Gordon equation (more on this in later chapters).

For a vector field $\phi_\mu(x)$ [$\mu = 0, 1, 2, 3$] for spin one particles with mass m, the Klein–Gordon equation is[1]

$$(\partial_\lambda \partial^\lambda + m^2)\phi_\mu(x) = 0$$

---

[1]Notations and the natural unit system are given in Appendices 1 and 2.

where

$$\partial_\lambda \partial^\lambda = \frac{\partial^2}{\partial t^2} - \nabla^2.$$

For the special case of mass zero particles, of spin one, the Klein–Gordon equation reduces to

$$\partial_\lambda \partial^\lambda \phi_\mu(x) = 0.$$

The classical wave equation for the electromagnetic four-vector potential $A_\mu(x)$, on the other hand, in the source-free region is

$$\partial_\lambda \partial^\lambda A_\mu(x) = 0.$$

An equation for a zero-mass particle of spin one (photon) in relativistic quantum mechanics turns out to be none other than the classical wave equation for the electromagnetic field of the 19th century that predates both relativity and quantum physics!

2

Lagrangian and Hamiltonian Dynamics

Lagrange's equations were formulated by the 18th century mathematician Joseph Louis Lagrange (1736–1813) in his book *Mathematique Analytique* published in 1788. In its original form Lagrange's equations made it possible to set up Newton's equations of motion, $F = dp/dt$, easily in terms of any set of generalized coordinates, that is, any set of variables capable of specifying the positions of all particles in the system. The generalized coordinates subsume the rectangular Cartesian coordinates, of course, but also include angular coordinates such as those in the plane polar or spherical polar coordinates. The generalized coordinates also allow us to deal easily with constraints of motion, such as a ball constrained to move always in contact with the interior surface of a hemisphere; the forces of constraints do not enter into the description of dynamics. As originally proposed, the Lagrange's equations provided a convenient way of implementing Newton's equations of motion.

Lagrange's equations became much more than just a powerful addition to the mathematical technique of mechanics when about 50 years later, in 1834, they became an integral part of Hamilton's principle of least action. Hamilton's principle represents the mechanical form of the calculus of variations that covers wide-ranging fields of physics. Lagrangian and Hamiltonian formulation of mechanics that established the basic pair of dynamical variables — position and momentum — is the precursor to the development of quantum mechanics and when it comes to the development of quantum field theory Lagrangian equations play an absolutely essential role.[1]

For a thorough discourse on the principle of least action in general, and the Hamilton's principle in particular, we will refer readers to many other excellent books on the subject. For our purpose we will focus on specific portions of the Lagrangian and Hamiltonian dynamics that describe the charged particles under the influence of an electromagnetic field. Not always fully appreciated, the Lagrangian and Hamiltonian descriptions of the electromagnetic interaction of the charged particles provide the foundation for quantum electrodynamics, and by extension, for the formulation of the quantum field theories of nuclear forces. The very origin of the field theoretical treatment of electromagnetic interaction traces its root to the classical Lagrangian and Hamiltonian dynamics.

The simplest way to show the equivalence of Lagrange's and Newton's equations is to use the rectangular coordinates, say, x_i ($i = 1, 2, 3$ for more conventional x, y, z). Using the notation $\dot{p} = dp/dt$ and $\dot{x} = dx/dt$, Newton's equations are

$$F_i = \dot{p}_i$$

$$p_i = m\dot{x}_i = \frac{\partial}{\partial \dot{x}_i}\left(\frac{1}{2}m\dot{x}_j^2\right) = \frac{\partial T}{\partial \dot{x}_i}$$

where T is the kinetic energy.

For a conservative system

$$F_i = -\frac{\partial V}{\partial x_i}$$

---

[1]Many excellent standard textbooks on classical mechanics include rich discussions on these subjects — Hamilton's principle, Lagrange's equations, and the calculus of variations. At the graduate level, the *de facto* standard on the subject is *Classical Mechanics* by Herbert Goldstein, Second edition, Addison-Wesley. At an undergraduate level, see, for example, *Classical Dynamics* by Jerry B. Marion, Second edition, Academic Press.

and Newton's equations are transcribed as

$$-\frac{\partial V}{\partial x_i} = \frac{d}{dt}\frac{\partial T}{\partial \dot{x}_i}.$$

In rectangular coordinates (and only in rectangular coordinates)

$$\frac{\partial T}{\partial x_i} = 0$$

and — this is an important point — for a conservative system

$$\frac{\partial V}{\partial \dot{x}_i} = 0.$$

Newton's equations can then be written as

$$\frac{\partial T}{\partial x_i} - \frac{\partial V}{\partial x_i} = \frac{d}{dt}\left(\frac{\partial T}{\partial \dot{x}_i} - \frac{\partial V}{\partial \dot{x}_i}\right)$$

which is Lagrange's equations, usually expressed as

$$\frac{d}{dt}\frac{\partial L}{\partial \dot{x}_i} - \frac{\partial L}{\partial x_i} = 0$$

where $L = T - V$ is the all-important Lagrangian function. The momentum p can be defined in terms of the Lagrangian function as

$$p_i = \frac{\partial L}{\partial \dot{x}_i}.$$

In terms of the generalized coordinates, denoted by q_i, that involve angular coordinates in addition to rectangular coordinates, the derivation of Lagrange's equations is slightly more involved. The terms $\partial T/\partial q_i$ are not zero, as in the case of rectangular coordinates, but are fictitious forces that appear because of the curvature of generalized coordinates. For example, in plane polar coordinates, where $T = (m/2)(\dot{r}^2 + r^2\dot{\theta}^2)$, we have $\partial T/\partial r = mr\dot{\theta}^2$, the centrifugal force. Lagrange's equation in terms of generalized coordinates remain in the same form, that is,[2]

$$\frac{d}{dt}\frac{\partial L}{\partial \dot{q}_i} - \frac{\partial L}{\partial q_i} = 0$$

with $L = T - V$ and momentum p is defined by

$$p_i = \frac{\partial L}{\partial \dot{q}_i}.$$

This definition of momentum p in terms of the Lagrangian represents a major extension of the original definition by Newton. In rectangular coordinates, it reduces to its original form, of course, but for those generalized coordinates corresponding to angles the new momentum corresponds to the angular momentum. In plane polar coordinates where $T = (m/2)(\dot{r}^2 + r^2\dot{\theta}^2)$, and $\partial V/\partial \dot{\theta} = 0$,

$$p_\theta = \frac{\partial L}{\partial \dot{\theta}} = \frac{\partial T}{\partial \dot{\theta}} = mr^2\dot{\theta}$$

which is the angular momentum corresponding to the angular coordinate.

This new definition of momentum is technically called the canonically conjugate momentum, that is, p_i being conjugate to the generalized coordinate q_i, and this pairing of (q_i, p_i) forms the very basis of the development of quantum mechanics and, by extension, the quantum field theory. Having thus become the standard basic dynamical variables, they are simply referred to as coordinates (dropping "generalized") and momenta (dropping "canonically conjugate").

This new definition of the canonically conjugate momentum, or simply momentum, has far-reaching consequences when the Lagrangian formulation is adopted to the case of charged particles interacting with the electromagnetic field. Often mentioned as a supplement within the framework of classical mechanics, this casting of Maxwell's equations into the framework of Lagrangian formulation

---

[2]Usually, Lagrange's equations are first derived from other physical principles — D'Alembert's principle or Hamilton's principle — and their equivalence to Newton's equations is shown to follow from the former. Here we follow the derivation as given in *Mechanics* by J.C. Slater and N.H. Frank, McGraw-Hill, which starts from Newton's equation, and then derive Lagrange's equations.

leads to non-mechanical extension of momentum and, as we will follow through in later chapters, provides the very foundation for the development of quantum electrodynamics.

Almost all forces we consider in mechanics are conservative forces, those that are functions only of positions, and certainly not functions of velocities, that is, $\partial V/\partial \dot{q}_i = 0$. There is, however, one very important case of a force that is velocity-dependent, namely, the Lorentz force on charged particles in electric and magnetic fields. In an amazing manner, the velocity-dependent Lorentz force fits perfectly into the Lagrangian formulation.

The Lagrangian equation can be written as

$$\frac{d}{dt}\frac{\partial T}{\partial \dot{q}_i} - \frac{\partial T}{\partial q_i} = -\frac{\partial V}{\partial q_i} + \frac{d}{dt}\frac{\partial V}{\partial \dot{q}_i}.$$

For conservative systems, $\partial V/\partial \dot{q}_i = 0$. For non-conservative systems when forces, and their potentials, are velocity-dependent, it is possible to retain Lagrange's equations provided that the velocity-dependent forces are derivable from velocity-dependent potentials — also called the generalized potentials — in specific form as required by Lagrange's equations, namely, the force is derivable from its potential by the recipe, expressed back in terms of the rectangular coordinates,

$$F_i = \left(-\frac{\partial}{\partial x_i} + \frac{d}{dt}\frac{\partial}{\partial \dot{x}_i} \right) V.$$

It is a rather stringent requirement and it turns out — very fortunate for the development of quantum electrodynamics — that the Lorentz force satisfies such requirement.

Putting $c = \hbar = 1$ in the natural unit system, the Lorentz force on a charge q in electric and magnetic fields, $\mathbf{E}$ and $\mathbf{B}$, is given by

$$\mathbf{F} = q\mathbf{E} + q(\mathbf{v} \times \mathbf{B})$$

where

$$\mathbf{E} = -\nabla\phi - \frac{\partial \mathbf{A}}{\partial t} \quad \text{and} \quad \mathbf{B} = \nabla \times \mathbf{A}$$

and ϕ and $\mathbf{A}$ are the scalar and vector potentials, respectively, defining the four-vector potential $A_\mu = (\phi, \mathbf{A})$. After some algebra,[3] the

---

[3]See Appendix 3.

Lorentz force can be expressed as

$$F_i = -\frac{\partial U}{\partial x_i} + \frac{d}{dt}\frac{\partial U}{\partial \dot{x}_i}$$

where

$$U = q\phi - q\mathbf{A} \cdot \mathbf{v}.$$

The Lagrangian for a charged particle in an electromagnetic field is thus

$$L = T - q\phi + q\mathbf{A} \cdot \mathbf{v}$$

and, as a result, the momentum — the new canonically conjugate momentum — becomes

$$\mathbf{p} = m\mathbf{v} + q\mathbf{A},$$

that is, mechanical Newtonian momentum plus an additional term involving the vector potential.

The Lagrangian formulation of mechanics was then followed by the Hamiltonian formulation based on treating the conjugate pairs of coordinates and momenta on an equal footing. This then led to the Poisson brackets for q's and p's and the Poisson brackets in turn led directly to quantum mechanics when they were replaced by commutators between the conjugate pairs of dynamical variables. For our purpose, we again focus on the motion of charged particles in an electromagnetic field.

In the Hamiltonian formulation, the total energy of a charged particle in an electromagnetic field is given by

$$E = \frac{1}{2m}(p_j - qA_j)(p_j - qA_j) + q\phi.$$

Comparing this expression to the total energy E for a free particle

$$E = \frac{1}{2m}p_j p_j$$

(p_j in each expression is the correct momentum for that case, that is, for free particles $m\dot{x}_j = p_j$, but for charged particles in an electromagnetic field $m\dot{x}_j = p_j - qA_j$), we arrive at the all-important substitution rule: the electromagnetic interaction of charged particles is given by replacing

$$E \Rightarrow E - q\phi$$

and

$$\mathbf{p} \Rightarrow \mathbf{p} - q\mathbf{A}.$$

In relativistic notations, this substitution rule becomes a compact expression

$$p^\mu \Rightarrow p^\mu - qA^\mu$$

where

$$p^\mu = (E, \mathbf{p}) \quad \text{and} \quad A^\mu = (\phi, \mathbf{A}).$$

As we shall see later, this substitution rule, obtained when Maxwell's equations for the electromagnetic field are cast into the framework of Lagrangian and Hamiltonian formulation of mechanics, is the very foundation for the development of quantum electrodynamics and, by extension, quantum field theory. It is *that* important. One must also note that whereas the Newtonian dynamics for particles went through modifications and extensions by Lagrange and Hamilton, the equations for the electromagnetic field not only remain unmodified but also, in fact, yielded a hidden treasure of instructions on how to incorporate the electromagnetic interaction.

3

Canonical Quantization

Transition from classical to quantum physics, together with the discovery of relativity of space and time, represents the beginning of an epoch in the history of physics, signaling the birth of modern physics of the 20th century. Quantum physics consists, broadly, of three main theories — non-relativistic quantum mechanics, relativistic quantum mechanics, and the quantum theory of fields. In each case, the principle of quantization itself is the same and it is rooted in the canonical formalism of the Lagrangian and Hamiltonian formulation of classical mechanics. In the Hamiltonian formulation, the coordinates and momenta are accorded an equal status as independent variables to describe a dynamical system, and this is the point of departure for quantum physics.

The two most important quantities in the Hamiltonian formulation is the Hamiltonian function and the Poisson bracket. The Hamiltonian function H — or just Hamiltonian, for short — is defined by

$$H(q, p, t) = \dot{q}_i p_i - L(q, \dot{q}, t)$$

where L is the Lagrangian. In all cases that we consider, $\dot{q}_i p_i$ is equal to twice the kinetic energy, T, and with the Lagrangian being equal to

$T - V$, the Hamiltonian corresponds to the total energy of a system, namely,

$$H = T + V.$$

The Poisson bracket of two functions u, v that are functions of the canonical variables q and p is defined as

$$\{u, v\} = \frac{\partial u}{\partial q_i} \frac{\partial v}{\partial p_i} - \frac{\partial u}{\partial p_i} \frac{\partial v}{\partial q_i}.$$

When u and v are q's and p's themselves, the resulting Poisson brackets are called the fundamental Poisson brackets and they are:

$$\{q_j, q_k\} = 0,$$
$$\{p_j, p_k\} = 0,$$

and

$$\{q_j, p_k\} = \delta_{jk}.$$

In terms of the Poisson brackets, the equations of motion for any functions of q and p can be expressed in a compact form. For some function u that is a function of the canonical variables and time, we have

$$\frac{du}{dt} = \{u, H\} + \frac{\partial u}{\partial t}$$

where $\{u, H\}$ is the Poisson bracket of $u(q, p, t)$ and the Hamiltonian H.

The transition from the Poisson bracket formulation of classical mechanics to the commutation relation version of quantum mechanics is affected by the formal correspondence ($\hbar$ is set equal to 1):

$$\{u, v\} \Rightarrow \frac{1}{i}[u, v]$$

where $[u, v]$ is the commutator defined by $[u, v] = uv - vu$, and on the left u, v are classical functions and on the right they are quantum mechanical operators. This transition from functions to operators and from the Poisson brackets to commutators is the very essence of quantization in a nut-shell.

In quantum mechanics, the time dependence of a system can be ascribed to either operators representing observables — momentum, energy, angular momentum and so on — or to wavefunctions representing the quantum state of a system. The former is called the Heisenberg picture and the latter Schrödinger picture (the third option is what is called the interaction picture in which both wavefunctions and operators are functions of time). In the Heisenberg picture, the equations of motion for any observable U is given by an exact counterpart of the classical equation in terms of the Poisson brackets, but with the Poisson bracket replaced by commutator, that is:

$$\frac{dU}{dt} = \frac{1}{i}[U, H] + \frac{\partial U}{\partial t}.$$

In the Schrödinger picture, operators representing observables are built up from those representing (canonically conjugate) momentum and energy by differential operators (expressed in rectangular coordinates),

$$p_j = -i\frac{\partial}{\partial x_j}$$

and since time t and $-H$ are also canonically conjugate to each other

$$E = i\frac{\partial}{\partial t}.$$

The wave equations of quantum mechanics, both non-relativistic and relativistic, are usually expressed in the Schrödinger picture and we have in the case of the non-relativistic quantum mechanics the time-dependent and time-independent Schrödinger's equations,

$$i\frac{\partial \psi(x_i, t)}{\partial t} = E\psi(x_i, t) \text{ (time-dependent)}$$

and from $E = p^2/(2m) + V$

$$\left(-\frac{1}{2m}\nabla^2 + V\right)\phi(x_i) = E\phi(x_i) \text{ (time-independent)}$$

where $\phi(x_i)$ is the space-dependent part of the total wavefunction $\psi(x_i, t)$. There are several wave equations in the relativistic quantum mechanics (Klein–Gordon, Dirac, Proca and other equations), but they must all first and foremost satisfy the relativistic energy–momentum relations

$$E^2 = p^2 + m^2$$

from which we obtain the Klein–Gordon equation

$$\left(\frac{\partial^2}{\partial t^2} - \nabla^2 + m^2 \right) \phi(x) = 0$$

or

$$(\partial_\lambda \partial^\lambda + m^2)\phi(x) = 0$$

that was mentioned in Chapter 1. As will be seen later, the quantum field theory is completely cast in the Heisenberg picture wherein the quantum mechanical wavefunctions themselves become operators.

The canonical procedure of quantization, be it non-relativistic quantum mechanics, relativistic quantum mechanics, or relativistic quantum field theory, can thus be compactly summarized as follows.

(i) First, find the Lagrangian function L which yields the correct equations of motion via the Lagrange's equation

$$\frac{d}{dt} \frac{\partial L}{\partial \dot{q}_i} - \frac{\partial L}{\partial q_i} = 0.$$

In the case of mechanical systems, $L = T - V$ and the equations of motion reduce to Newton's equations of motion.

(ii) Define canonically conjugate momentum p with the help of L,

$$p_i = \frac{\partial L}{\partial \dot{q}_i}.$$

(iii) Quantization is effected when we impose the basic commutation relations

$$[q_j, q_k] = [p_j, p_k] = 0$$

and

$$[q_j, p_k] = i\delta_{jk}.$$

(iv) The wave equations in quantum mechanics, both non-relativistic and relativistic, are obtained, in the Schrödinger picture, by the operator representation of momentum and energy (expressed in rectangular coordinates) as

$$p_j = -i\frac{\partial}{\partial x_j} \quad \text{and} \quad E = i\frac{\partial}{\partial t}.$$

As will be seen later, in the quantum field theory the very quantum mechanical wavefunctions themselves become operators for generalized coordinates and the corresponding canonically conjugate momenta are defined by the same recipe via the Lagrangian. The Lagrangian function thus is an absolutely essential element in any quantum physics, be it quantum mechanics or quantum field theory.

4

Particles and Fields II: Duality

The departure of quantum mechanics from classical mechanics is quite drastic, rather extreme in contemporary parlance. Ordinary physical quantities are replaced by quantum mechanical operators that do not necessarily commute with each other and the Heisenberg's uncertainty principles between the canonically conjugate pairs of variables, between coordinates and momenta and between time and energy, deny the complete determinability of classical physics.

The most basic and defining characteristic of quantum mechanics — often called the central mystery of quantum mechanics — is the uniquely dual nature of matter called the wave–particle duality. In the microscopic scale of quantum world — of atoms, nuclei and elementary particles — a physical object behaves in such a way that exhibits the properties of both a wave and a particle. Often the wave–particle duality of quantum world is presented as physical objects that are *both* a wave and a particle. To us, with human intuition being nurtured in the macroscopic world, this simplistic picture of wave–particle duality is, of course, completely counterintuitive.

A quantum mechanical object is actually neither a wave in the classical sense nor a particle in the classical sense, but rather it defines a totally new reality, the quantum reality, that in some circumstances exhibits properties much like those of classical particles and in some other circumstances displays properties much like those of classical wave. The new quantum reality can be stated as being "neither a wave nor a particle but is something that can act sometimes much like a wave and at other times much like a particle." The new quantum reality, the wave–particle duality, thus combines the classical dichotomy of particles and fields, waves being specific examples of a field broadly defined as an entity with spatial extension.

Let us briefly recapitulate what is meant by a particle in the classical sense. First, it has mass and occupies one geometric point in space; that is, it has no spatial extension. When it moves, under the influence of a force, it moves from one point at one time to another point at another moment in time. The entire trace of its motion is called its trajectory. Once the initial position and velocity are fixed, Newton's equations of motion determine completely its trajectory. If the laws of motion dictate a particle to pass through a particular position A at some time t, the particle will pass through that point. There is no way the particle can be seen to be passing through any other positions at that same time. It will pass through the position A and nowhere else. Furthermore, without any force to alter its course, a particle cannot simple decide to change its direction of motion and can go to other positions. That is a no–no.

Another defining characteristic of a particle in the classical sense is the way in which it impacts, that is, how it interacts with another object. The classical particle interacts with others at a point of collision; some of its energy and momentum are transferred to others at that point of impact. It is the point-to-point transfer of energy and momentum that is the basic dynamical definition of what a particle is in the classical sense.

We can easily contradistinguish the kinematical and dynamical aspects of a classical wave from those of a classical particle. First and foremost, a wave is certainly not something that is defined at a geometrical point. On the contrary, a wave is definitely an extended object — a wave train with certain wavelength and frequency — and furthermore it does not travel along a point-to-point trajectory, but rather propagates in all directions. A sound or light wave propagates from its source in expanding spheres in all directions. In a room whose walls are shaped as the interior surface of a sphere, a wave will hit all points of the wall at the same time.

As a wave also carries its own energy and momentum, the way it propagates in all directions dictates the way it transfers energy and momentum, everywhere in all directions as it comes into contact with other matter. There is no point-to-point transfer as far as the wave is concerned. The contrast between the classical particle and wave could not have been more diagonally opposite, and this is what the new quantum reality called wave–particle duality brings together!

Now, an important caveat is in order about a matter of terminology in quantum physics. The new quantum reality, the wave–particle duality, describes a quantum *thing* that is neither a particle in classical sense nor a wave in classical sense. We can shorten the name to simply *duality*, that is, electrons, protons, neutrons and photons, etc should all be called *duality*, certainly not *particle* nor *wave*. Our reluctance or inability to part with the word "particle" is such that, however, the objects in the quantum world — be they electrons, photons, protons, neutrons, quarks and whatever — are continually referred to as "particles," as in unstable particles, elementary particle physics and so on. What has happened is that the meaning of the word "particle" has gone through a metamorphosis: the word particle when applied to entities in the quantum world actually means duality, the wave–particle duality. Terminologies have gone from classical wave and classical particle to wave–particle duality, or just duality for short, and the "duality" has morphed back into "particle." We will conform to this practice and from this point on in this book, the word "particle" will stand for duality and the word "particle" in classical sense will always be referred to as, "particle in the classical sense." This is a somewhat confusing story of the evolution in the meaning of the word "particle."

The properties of this (new quantum) particle are thus this. When it impacts, that is, interacts with other particles, it behaves much as the way a classical particle does, that is, a transfer of energy and momentum occurs at a point. But it travels in all directions like a classical wave. Since the impact occurs at a point, the question then arises as to what determines the particle to impact at one point at some time and at a different point at another time. In other words, what determines its preference to land at a particular point, and not elsewhere, at one time and land at another point at another time. This is the crux of the matter of quantum mechanics: the particle carries with it the information that determines the probability of its landing at a particular point.

Going back to the example of a room with the walls shaped like the interior walls of a sphere, the (quantum) particle can strike anywhere on the wall, impacting a particular point as if it were a classical particle (making a point mark on the wall). If you repeat the experiment again and again, the particle will land at points (one at a time) all over the wall, but with varying probabilities, at some points more often than at some other points.

The defining properties of the (quantum) particle can thus be summarized as:

(i) It spreads like a classical wave, in all directions.
(ii) However, it impacts like a classical particle.
(iii) It carries with it its own information on the probability of where it is likely to impact.

The next logical question then is what determines its probability. And this is what the equations of quantum mechanics, such as the Schrödinger's equation of non-relativistic quantum mechanics, provide as their solutions, namely, wavefunctions.

This is phase II in the evolution of particles and fields, first the classical dichotomy and now the quantum duality.

Equations for Duality

The wavefunctions for a particle (in the new sense of wave–particle duality) are to be determined as solutions of quantum mechanical wave equations and these wavefunctions provide information on the probability of the particle impacting at or near a particular position. We have already mentioned the quantum mechanical differential operators corresponding to momentum and energy and by substituting these operator expressions to either the non-relativistic formula for total energy or the relativistic one, we obtain the corresponding equations of quantum mechanics.

The Schrödinger's Equation

In non-relativistic quantum mechanics, the equation in question is the Schrödinger's equation, which is the central and only wave equation for non-relativistic quantum mechanics. As mentioned in Chapter 3, the time-dependent Schrödinger's equation is

$$i\frac{\partial \psi(x_i, t)}{\partial t} = E\psi(x_i, t).$$

Writing $\psi(x_i, t) = \phi(x_i)T(t)$, the function of time only has solutions in the form of

$$T(t) = \exp(-iEt)$$

where E is the quantized values (eigenvalues) of energy determined from the time-independent Schrödinger's equation obtained from

$$\left(-\frac{1}{2m}\nabla^2 + V\right)\phi(x_i) = E\phi(x_i).$$

The absolute square of the solutions $|\phi(x_i)|^2$, for each allowed values of E, is the probability distribution function of finding the particle in question in a state with a particular value of energy, E, in a small region between x and $x + dx$. The wavefunctions are referred to as the probability amplitudes, or just amplitudes, and the absolute square of wavefunctions as the probability density, or just probability. This interpretation, the postulate of probabilistic interpretation of wavefunctions, is one of the basic tenets of quantum mechanics and is the "heart and soul" of wave–particle duality.

The Schrödinger's equation and its solutions, however, fall short of accommodating one of the basic attributes of particles, the spin of a particle. Since the spin is an intrinsic property of a particle that is not at all associated with the spatial and temporal coordinates of the particle, it is one of the internal degrees of freedom of a particle — as opposed to the spatial and temporal coordinates being the external degrees of freedom — and the Schrödinger's equation is not set up to deal with any such internal degrees of freedom. The electric charge of a particle is another example of internal degrees of freedom that has nothing to do with the spatial and temporal coordinates.

In the case of electronic orbits of an atom, for example, the solutions $\phi(x_i)$ successfully specify the radii of the orbits (the principal quantum number), the angular momentum of an electron in an orbit (the total angular momentum quantum number), and the tilts of the planes of an orbit (the magnetic quantum number). The complete knowledge of the structure and physical properties of atoms, however, requires specification of the electron spin and Pauli's exclusion principle, without which the physics of atoms, and by extension, all known matter in the universe, would not have been what it is. In this sense, while it is the enormously successful central equation for atomic physics, the Schrödinger's equation falls short of completing the story of atoms. The spin part of information is simply tacked onto the wavefunctions as an add-on, in the case of electrons, by a two-component (for spin-up and spin-down) one-column matrices.

The spin of a particle finds its rightful place only when we proceed to relativistic quantum mechanics. Particles with half-integer spin — generically called the fermions — such as electrons, protons and neutrons that constitute all known matter obey the relativistic wave equation called the Dirac equation (see below), wherein only the total angular momentum defined as the sum of orbital angular momentum and spin is conserved, whereas in the non-relativistic case the conserved quantity is the orbital angular momentum only. The wave equations for relativistic quantum mechanics are obtained from the relativistic energy momentum relation by an operator substitution

$$p^\mu = i\partial^\mu = i\frac{\partial}{\partial x_\mu} = i\left(\frac{\partial}{\partial t}, -\nabla\right)$$

into

$$E^2 - p^2 = m^2 \quad \text{or} \quad p^\mu p_\mu = m^2.$$

The Klein–Gordon Equation

Particles with spin zero, those with no spin at all, are described by a scalar amplitude, $\phi(x)$, that is invariant under the Lorentz transformation, meaning that the amplitude remains the same as observed in any inertial frame. For brevity, we will use the notation x for space-time coordinate four-vector (Appendix 2, Notations). From $p^\mu p_\mu = m^2$, we have the Klein–Gordon equation

$$(\partial^\mu \partial_\mu + m^2)\phi(x) = 0.$$

The Schrödinger wavefuntion is also a scalar wavefunction; it does not address the spin degrees of freedom. For particles of other values of spin, spin one for vector bosons and spin one-half for fermions, the wavefunctions are not scalars; they are four-vector wavefunctions for spin one vector bosons and four-component spinors for fermions, and each satisfies its own set of equations over and beyond the Klein–Gordon equation. But any relativistic wavefunction, regardless of the spin of the particle, must first and foremost satisfy the Klein–Gordon equation.

The Dirac Equation

The Dirac equation is the most significant achievement of relativistic quantum mechanics. It successfully incorporated the spin of a particle as the necessary part of the particle's total angular momentum, and it also predicted the existence of antiparticles — positrons, antiprotons, antineutrons, and so forth. Since 99.9% of the known matter in the universe is made up of electrons, protons and neutrons, all of which are spin one-half fermions, the Dirac equation applies to the basic particles that make up all known matter. One can go so far as to claim that the Dirac equation and relativistic quantum mechanics are virtually synonymous.

What originally prompted Dirac to search for and discover the Dirac equation is simple and straightforward enough. The Klein–Gordon equation is a second-order differential equation — second derivatives with respect to both space and time — and as a relativistic equation for single particle, it encounters some difficulties; the nature of second-order differential equations and the probability interpretation of quantum mechanics clash. (We will not discuss these difficulties here, but mention that difficulties do arise for the Klein–Gordon equation as one-particle equation becomes resolved when solutions of the Klein–Gordon equation are treated as quantized fields in quantum field theory.) Rather than a second-order equation, Dirac wanted a first-order linear equation containing only the first derivatives with respect to both space and time, that is, linear with respect to four-vector derivates.

The process of going from a second-order expression to a first-order one is a matter of factorization and let us dwell on this matter here. The simplest algebraic factorization is, of course, the factorization of $x^2 - y^2$:

$$x^2 - y^2 = (x + y)(x - y).$$

Factorization of $x^2 + y^2$, however, cannot be done in terms of real numbers but needs the help of complex numbers:

$$x^2 + y^2 = (x + iy)(x - iy).$$

Factorization of a three-term expression such as $x^2 + y^2 + z^2$ requires much more than just numbers, real or complex; and we must rely on matrices. Consider the three Pauli spin matrices σ, given in their standard representation as

$$\sigma_x = \begin{pmatrix} 0 & 1 \\ 1 & 0 \end{pmatrix}, \quad \sigma_y = \begin{pmatrix} 0 & -i \\ i & 0 \end{pmatrix}, \quad \sigma_z = \begin{pmatrix} 1 & 0 \\ 0 & -1 \end{pmatrix},$$

satisfying the anticommutation relations

$$\{\sigma_j, \sigma_k\} \equiv \sigma_j \sigma_k + \sigma_k \sigma_j = 2\delta_{jk}.$$

For any two vectors $\mathbf{A}$ and $\mathbf{B}$ that commute with σ, we have the following identity

$$(\sigma \cdot \mathbf{A})(\sigma \cdot \mathbf{B}) = \mathbf{A} \cdot \mathbf{B} + i\sigma \cdot (\mathbf{A} \times \mathbf{B}).$$

When applied to only one vector, the identity reduces to

$$(\sigma \cdot \mathbf{A})(\sigma \cdot \mathbf{A}) = \mathbf{A} \cdot \mathbf{A}$$

and this allows, in terms of 2×2 anticommuting matrices, factorization of three-term expressions, such as

$$p_x^2 + p_y^2 + p_z^2 = \mathbf{p} \cdot \mathbf{p} = (\sigma \cdot \mathbf{p})(\sigma \cdot \mathbf{p}).$$

Now we can factorize $p^\mu p_\mu = E^2 - (p_x^2 + p_y^2 + p_z^2)$, a four-term expression is thus

$$p^\mu p_\mu = E^2 - (p_x^2 + p_y^2 + p_z^2) = E^2 - (\sigma \cdot \mathbf{p})(\sigma \cdot \mathbf{p})$$
$$= (E + \sigma \cdot \mathbf{p})(E - \sigma \cdot \mathbf{p}).$$

This has led to the relativistic wave equation for massless fermions in the form of

$$p^\mu p_\mu \varphi_\alpha = (E + \sigma \cdot \mathbf{p})(E - \sigma \cdot \mathbf{p})\varphi_\alpha = 0 \quad \text{with } \alpha = 1, 2,$$

where φ_α is a two-component wavefunction (since the Pauli matrices are 2×2 matrices). This used to be the wave equation for two-component zero-mass electron neutrinos (nowadays, the neutrinos are considered to have mass, however minute it may be). Factorization by the use of three 2×2 matrices renders the amplitude $\varphi_\alpha(x)$ to be a two-component column matrix, called a spinor.

This then brings us to the Dirac equation as a result of factorizing the five-term expression of the Klein–Gordon equation, $p^\mu p_\mu - m^2$. It cannot be brought to a linear equation even with the help of three 2×2 Pauli matrices. Dirac has shown that factorization is possible but only with the help of four 4×4 matrices that are built up from the 2×2 Pauli matrices. Such is the humble beginning of the Dirac equation that came to govern the behavior of all particles of half-integer spin. The five-term expression can be factorized thus

$$p^\mu p_\mu - m^2 = E^2 - (p_x^2 + p_y^2 + p_z^2) - m^2$$
$$= (\gamma^\mu p_\mu + m)(\gamma^\nu p_\nu - m)$$

where the four γ^μ matrices are required to satisfy the anticommutation relations

$$\gamma^\mu \gamma^\nu + \gamma^\nu \gamma^\mu = 2g^{\mu\nu}$$

and by virtue of which

$$\gamma^\mu p_\mu \gamma^\nu p_\nu = p^\mu p_\mu.$$

It is the four-dimensional analogue of the three-dimensional relations

$$(\sigma \cdot \mathbf{p})(\sigma \cdot \mathbf{p}) = \mathbf{p} \cdot \mathbf{p}.$$

Of the many different matrix representations of four γ matrices, the most-often used is where

$$\gamma^0 = \begin{pmatrix} I & 0 \\ 0 & -I \end{pmatrix}, \quad \text{and} \quad \gamma^k = \begin{pmatrix} 0 & \sigma^k \\ -\sigma^k & 0 \end{pmatrix}$$

and the σ's are Pauli's spin matrices and I is the 2×2 unit matrix.

The Dirac equation then becomes, replacing p_μ by $i\partial_\mu$,

$$(i\gamma^\mu \partial_\mu - m)\psi_\alpha(x) = 0.$$

The Dirac amplitude $\psi_\alpha(x)$ with $\alpha = 1,2,3,4$ is now a four-component spinor which, in the standard representation, consists of positive-energy solutions with spin up and down and negative-energy solutions with spin up and down.

It is clear from the factorization of the second-order relativistic energy momentum relation that each component of the Dirac amplitude must also independently satisfy the Klein–Gordon equation, that is,

$$(\partial^\mu \partial_\mu + m^2)\psi_\alpha(x) = 0 \quad \text{for each } \alpha = 1, 2, 3, 4.$$

The Dirac equation imposes further conditions over and beyond the Klein–Gordon equation — very stringent interrelations among the components and their first derivatives — among the four components of the solution. This can be seen when the Dirac equation is fully written out in 4×4 matrix format using an explicit representation of γ-matrices such as shown above.

The Dirac equation is the centerpiece of relativistic quantum mechanics. All textbooks on the subject devote a substantial amount of the contents to detailing all aspects of this equation — proof of its relativistic covariance, the algebraic properties of Dirac matrices, as γ matrices are called, the bilinear covariants built from its four-component solutions, and many others — and, in fact, virtually all textbooks on quantum field theory also include extensive discussions about the equation, before embarking on the subject of field quantization. We will not discuss here the extensive properties of Dirac equation, but suffice it to say that the equation is perhaps the most important one in all of quantum mechanics. It is an absolutely essential tool in elementary particle physics. After all, it is the equation for all particles that constitute the known matter in the universe — all fermions of spin one-half which also includes all leptons, of which the electron is the premier member, and all quarks, out of which such particles as protons and neutrons are made. (More on leptons and quarks in later chapters.)

At this point, let us briefly recap what we traced out in the previous five chapters, including this one. The evolution in our treatment of material particles has come through several phases. The abstract concept of a point mass in Newtonian mechanics remained intact through the development of Lagrangian and Hamiltonian dynamics. When quantum mechanics replaced the classical dynamics of Newton, Lagrange and Hamilton, the concept of particle went through a fundamental revision, from that of a well-defined classical point mass to one of quantum-mechanical wave–particle duality in which it is neither a particle in the classical sense nor a wave in the classical sense, but a new reality in the quantum world that displays both particle-like and wave-like properties. In non-relativistic quantum mechanics, the probability amplitude for this wave–particle duality is to be determined as solutions of the Schrödinger's equation and in the fully relativistic case as solutions of the Dirac equation. Prior to the development of quantum field theory, the evolution in the concept of particle consisted essentially of two stages: first, Newton's point-mass and then the quantum-mechanical wave-particle duality. This concept of particles would then go through a radical change within the framework of quantum field theory.

One might notice at this point as to why not a single word has been mentioned of the wave equations for electromagnetic fields, which would lead to the equation for photons, the equation that along with the Dirac equation for fundamental fermions completes the founding pillars of quantum field theory. It has not been included up to this point for a very good reason: the wave equation for the electromagnetic field is an equation not of quantum mechanics but of classical physics. The equation for the electromagnetic field predates the advent of both relativity and quantum mechanics. One might go so far as to say that the equation for wave, the electromagnetic wave, has been "waiting" all this while for the equations for particles to "catch up" with it! We will now turn to this classical wave equation for the electromagnetic field in the next chapter.

6

Electromagnetic Field

The classical theory of electromagnetism, as mentioned in Chapter 1, developed along an entirely different path from that of Newton's classical mechanics. From day one, electromagnetism was based on properties of force fields — the electric and magnetic fields that are extended in space. An electric field due to a point charge, for example, is defined over the entire three-dimensional space surrounding the point charge. The works of Coulomb, Gauss, Biot–Savart, Ampère, and Faraday led Maxwell to the great unification of electricity and magnetism into a single theory of an electromagnetic field. Together with Einstein's theory of gravitational field, Maxwell's theory of electromagnetic field is one of the most elegant of classical field theories.

Maxwell's equations are given, in the natural unit system, as

$$\nabla \cdot \mathbf{E} = \rho, \qquad \nabla \times \mathbf{B} - \frac{\partial \mathbf{E}}{\partial t} = \mathbf{J}, \quad \text{(inhomogeneous)}$$

$$\nabla \cdot \mathbf{B} = 0, \qquad \nabla \times \mathbf{E} + \frac{\partial \mathbf{B}}{\partial t} = 0. \quad \text{(homogeneous)}$$

where $\mathbf{E}$ and $\mathbf{B}$ are the electric and magnetic fields and ρ and $\mathbf{J}$ are the electric charge and current densities. The electric and magnetic fields can be expressed in terms of a scalar potential ϕ and a vector potential $\mathbf{A}$ as

$$\mathbf{B} = \nabla \times \mathbf{A}, \qquad \mathbf{E} = -\nabla\phi - \frac{\partial \mathbf{A}}{\partial t},$$

and the two homogeneous Maxwell's equations are satisfied identically.

The electric charge and current densities ρ and $\mathbf{J}$ are components of a single four-vector $J^\mu = (\rho, \mathbf{J})$ and likewise the scalar and vector potentials ϕ and $\mathbf{A}$ are components of a four-vector potential $A^\mu = (\phi, \mathbf{A})$. The electric and magnetic fields, $\mathbf{E}$ and $\mathbf{B}$, correspond to components of the antisymmetric electromagnetic field tensor $F^{\mu\nu}$ defined as

$$F^{\mu\nu} \equiv \partial^\mu A^\nu - \partial^\nu A^\mu = \begin{pmatrix} 0 & -E_x & -E_y & -E_z \\ E_x & 0 & -B_z & B_y \\ E_y & B_z & 0 & -B_x \\ E_z & -B_y & B_x & 0 \end{pmatrix}.$$

The electromagnetic field tensor is thus a four-dimensional "curl" of the four-vector potential. In terms of the electromagnetic field tensor, the inhomogeneous Maxwell's equations become

$$\partial_\mu F^{\mu\nu} = J^\nu.$$

We can now draw two very important conclusions about Maxwell's equations. First, the four-potential $A^\mu = (\phi, \mathbf{A})$ is not unique in the sense that the same electromagnetic field tensor $F^{\mu\nu}$ is obtained from the potential

$$A^\mu + \partial^\mu \Lambda = \left(\phi + \frac{\partial \Lambda}{\partial t}, \mathbf{A} - \nabla\Lambda\right),$$

where $\Lambda(x)$ is an arbitrary function and its contribution to $F^{\mu\nu}$ is identically zero (it is the four-dimensional analogue of the curl of gradient being identically zero). This freedom to shift the four-potential $A^\mu = (\phi, \mathbf{A})$ by the four-gradient of an arbitrary function is called *gauge transformation* and it forms the basis for the quantum field theory for the standard model, sometimes also called the theory of gauge fields.

The second conclusion is no less important. The source-free ($J^\mu = 0$) inhomogeneous Maxwell's equations are

$$\partial_\mu F^{\mu\nu} = \partial_\mu \partial^\mu A^\nu - \partial^\nu \partial_\mu A^\mu = 0.$$

Using the freedom of gauge transformation, we can set $\partial_\mu A^\mu = 0$. The choice of the arbitrary function $\Lambda(x)$ to render $\partial_\mu A^\mu$ as always being zero is referred to as the Lorentz gauge. With such an option, Maxwell's equations reduce to

$$\partial_\mu \partial^\mu A^\nu = 0,$$

which, as mentioned in Chapter 1, is exactly the zero-mass case of Klein–Gordon equation.

At the risk of being repetitive, let us emphasize this remarkable point that Maxwell's equations are classical wave equations for the four-potential, and they predate both relativity and quantum mechanics. In this amazing twist, a window has opened up for us to look at the relativistic quantum mechanical wave equations, such as the Klein–Gordon and Dirac equations, in an entirely new light.

7

Emulation of Light I: Matter Fields

We are now at the point, after the first six chapters, to look back and compare where the equations of motion for the field and particles stand with respect to each other. As far as electromagnetic fields are concerned, the equations remain intact in its original form, as Maxwell had written down. As discussed in the last chapter, Maxwell's equations for the radiation field, that is, in the source-free region, arc of very compact expression. In terms of the four-vector potential and for a particular choice of gauge called the Lorentz gauge, the equations are expressed as

$$\partial_\mu \partial^\mu A^\nu = 0$$

which also coincided with the Klein–Gordon equation for mass-zero case. The equations for particles, on the other hand, evolved through several phases — from Newton to Lagrange and Hamilton and through relativity and quantum mechanics — and ended up in the form of wave equations for non-interacting one-particle within the framework of relativistic quantum mechanics, the Klein–Gordon and Dirac equations being prime examples.

Relativistic quantum mechanical equations as one-particle equations, however, suffer from some serious interpretative problems. For example, the Klein–Gordon equation could not avoid the problem of occurrence of negative probabilities while the Dirac equation suffered from the appearance of negative-energy levels. A new insight was definitely required to proceed to the next phase in the evolution of theories of particles. Such insight would come from the quantization of the electromagnetic field. We will discuss the formalism of the quantization of classical fields in the next chapter. Suffice it to say here that when the radiation field (the electromagnetic field in the source-free region) was quantized, following the recipe for canonical quantization the quantal structure of such quantized radiation field corresponded to photons of Planck and Einstein, the particles of light. This point needs to be repeated: *When a classical field is quantized (in the manner as will be discussed in the next chapter), the quanta of the field are the particles represented by the classical field equation.* This relationship between the classical electromagnetic field and photons, the discreet energy quanta of the radiation field that correspond to particles of light with no mass, provided an entirely new insight into the interpretation of particles. The concept of particles would then go through another fundamental evolution, from that of quantum-mechanical wave–particle duality to that of the quanta of a quantized field.

For particles to be described by relativistic quantum fields, however, there were no corresponding classical fields. We know of only two classical fields in nature, the electromagnetic and gravitational fields. Where and how do we find the classical fields whose quanta correspond to particles satisfying the Klein–Gordon or the Dirac equations? And it is here that we find one of the fundamental conceptual shifts needed to proceed to the next level. *Relativistic quantum-mechanical wave equations such as the Klein–Gordon and Dirac equations are to be reinterpreted as classical field equations at the same level as Maxwell's equation for the classical electromagnetic field!* This is definitely a leap of faith.

Overnight the wave amplitudes for particles (that is, the particle–wave dualities) were turned into corresponding classical fields and the wave equations of relativistic quantum mechanics were turned into corresponding "classical" equations for the classical fields. No equations were modified and all notations remained intact. The wave amplitude $\varphi(x)$ became the classical field $\varphi(x)$ and relativistic quantum-mechanical equations became wave equations for classical fields. This turned out to be one of the most subtle conceptual switches in the history of physics. This was the first instance — it would not be the last — in which matter emulated radiation. This is precisely how we arrived, in the early days of 1930s and 1940s, at the very beginning of quantum field theory of matter — equations for matter simply emulating those for radiation. So, at this point, every wave equation for matter as well as radiation is a classical wave equation for classical fields, some real (Maxwell's equations) and others "imitations" (relativistic quantum mechanical wave equations). We have the truly classical field of the electromagnetic field satisfying

$$\partial_\mu \partial^\mu A^\nu = 0,$$

and the "imitation" classical fields, which are the redressed relativistic quantum-mechanical wave equations, satisfying equations such as those of Klein–Gordon and Dirac, *but now viewed from this point forward as classical field equations*:

$$(\partial^\mu \partial_\mu + m^2)\phi(x) = 0$$

and

$$(i\gamma^\mu \partial_\mu - m)\psi_\alpha(x) = 0.$$

Strictly speaking, the classical Klein–Gordon or Dirac field does not exist in the macroscopic scale. No signals are transmitted by these "fields" from one point to another in space in the same way radio signals are carried by the classical radiation field. Reinterpretation of these relativistic quantum-mechanical wave equations as classical field equations is the first preliminary step toward establishing the quantum field theory of matter particles. Once these "imitation" classical fields are quantized in exactly the same manner as the electromagnetic field, the resulting theory of matter particles interacting with photons — quantum electrodynamics — turned out to be the most successful theory for elementary particles to date. In this sense, the redressing of relativistic quantum-mechanical wave equations into "imitation" classical field equations is one of many examples of "the end justifying the means." That is its rationalization.

Road Map for Field Quantization

We are now ready to proceed with the quantization of classical fields — the classical electromagnetic, Klein–Gordon and Dirac fields — that were discussed in the last chapter. The quantization of these fields is to be carried out following the rules of canonical quantization, as discussed in Chapter 3. Before imposing canonical quantization onto the classical fields, however, we need to extend the Lagrangian formalism from that of point mechanics to one more suitable for continuous classical field variables.

First and foremost is the question of generalized coordinates . The generalized coordinates $q_i(t)$ with discrete index $i = 1, 2, \ldots, n$ for a system with n degrees of freedom is taken to the limit $n \to \infty$ and in that limit the values of a field at each point of space are to be considered as independent generalized coordinates. Consider a simple mechanical example of a continuous string: the vertical displacement function, say, $\rho(\mathbf{x}, t)$, stands for the amplitude of displacements of the continuous string at position $\mathbf{x}$ and at time t and its values at each position can be taken as independent generalized coordinates. The discrete index i of the generalized coordinates for point mechanics is replaced by the continuous coordinate variable $\mathbf{x}$, and the fields themselves — $A_\mu(x)$ of the electromagnetic field, $\phi(x)$ of the Klein–Gordon field, $\psi_\alpha(x)$ of the Dirac field, and so on — take the place of generalized coordinates.

The canonical formalism for fields requires the canonically conjugate momenta that are to be paired with field variables, and the momenta that canonically conjugate to fields can be defined in terms of the Lagrangian that yields correct equations of motion via Lagrange's equations of motion. In Chapter 2, we took the simplest approach to obtaining Lagrange's equation, starting from Newton's equations of motion. There is another way of obtaining Lagrange's equations that is more formal than the direct approach we took in Chapter 2, and that is to derive Lagrange's equation from what is called Hamilton's principle of least action for particle mechanics. The resulting solution of Hamilton's principle is known mathematically as the Euler equation and Lagrange's equation is a specific example of this more generic Euler equation adopted for particle mechanics. Often, for this reason, Lagrange's equations are also referred to as the Euler–Lagrange equations. We will not get into the details of this formalism here, especially since all that we really need is the expression for Lagrangian that will help define expressions for the momenta canonically conjugate to fields.

For classical fields, it is more convenient to use the Lagrangian densities $\mathcal{L}$ defined as

$$L \equiv \int_{-\infty}^{\infty} d^3x\, \mathcal{L}\left(\phi, \frac{\partial\phi}{\partial x^\mu}\right)$$

and the Euler–Lagrange equations in terms of the Lagrangian densities are given as

$$\frac{\partial}{\partial x^\mu} \frac{\partial \mathcal{L}}{\partial(\partial\phi/\partial x^\mu)} - \frac{\partial \mathcal{L}}{\partial \phi} = 0.$$

Comparing the Euler–Lagrange equations above with the Lagrange's equations given in Chapter 2, we note that the only change is in the leading term where derivatives with respect to time only are replaced by derivatives with respect to all four space-time coordinates x^μ. The momenta conjugate to a field are defined via the Lagrangian density in much the same way as for the case of particle mechanics, thus:

$$\pi(\mathbf{x}, t) \equiv \frac{\partial \mathcal{L}}{\partial(\partial\phi/\partial t)}.$$

We can now state in one paragraph what the quantization of classical fields is all about: (1) start with the classical field equations, Maxwell's, Klein–Gordon, and Dirac equations for radiation field and matter fields, respectively, (2) seek a Lagrangian density for each field that reproduces the field equations via the Euler–Lagrange equations (this is about the only use of Euler–Lagrange equations in this context), (3) with the help of these Lagrangian densities, define momenta canonically conjugate to the fields, and (4) carry out the quantization by imposing commutation relations on these canonically conjugate pairs. After imposing quantization, the fields and their momenta become quantum mechanical operators. *That* sums up in a nutshell what the quantum field theory is all about.

As the fields and their conjugate momenta are both functions of time, as well as of space, they become, upon quantization, operators that are functions of time and this necessarily casts the whole quantum field theory in the Heisenberg picture of quantum theory. As mentioned briefly in Chapter 3, there are two equivalent ways in which the time development of a system can be ascribed to: either operators representing observables or states represented by time-dependent wavefunctions. The former is called the Heisenberg picture and the latter Schrödinger picture. In one-particle quantum mechanics, both non-relativistic and relativistic, it is usually more convenient to adopt the Schrödinger picture and the time development of a system is given by the wavefunctions as functions of time. In quantum field theory wherein the time-dependent fields and their conjugate momenta become operators, the formalism is necessarily in the Heisenberg picture. Let us spell out the bare essence of the relationship between the two pictures, as far as canonical quantization rules are concerned.

In the Schrödinger picture, the wavefunction $\psi(t)$ carries the time development information, that is, if the initial state at an arbitrary time, say $t = 0$, is specified, the Schrödinger's equation determines the state at all future times. The commutation relations for the canonical pairs of operators are, as given before,

$$[q_j, p_k] = i\delta_{jk}$$
$$[q_j, q_k] = [p_j, p_k] = 0.$$

In the Heisenberg picture, the wavefunction is time-independent and is related to that of the Schrödinger picture by

$$\psi_{\mathbf{H}} \equiv \psi_{\mathbf{S}}(0)$$

and the commutators between q and p become, for an arbitrary time t,

$$[q_j(t), p_k(t)] = i\delta_{jk}$$
$$[q_j(t), q_k(t)] = [p_j(t), p_k(t)] = 0.$$

In the continuum language of fields and conjugate momenta, they become

$$[\phi(\mathbf{x}, t), \phi(\mathbf{x}', t)] = [\pi(\mathbf{x}, t), \pi(\mathbf{x}', t)] = 0$$
$$[\phi(\mathbf{x}, t), \pi(\mathbf{x}', t)] = i\delta^3(\mathbf{x} - \mathbf{x}')$$

where the Dirac delta function replaces δ_{jk} and is defined by

$$\int d^3x\, \delta^3(\mathbf{x} - \mathbf{x}') f(\mathbf{x}') = f(\mathbf{x}).$$

In sum, four items — field equations, the Lagrangian densities, momenta conjugate to the fields, and the commutation relations imposed on them — provide the basis of what is called the quantum field theory.

   The determination of the Lagrangian density thus plays the very starting point for quantum field theory and Lagrangian densities are so chosen such that the Euler–Lagrange equations reproduce the correct field equations for a given field. The choice, however, is not unique since the Euler–Lagrange equations involve only the derivatives of the Lagrangian densities. A Lagrangian density is chosen to be the simplest choice possible that meets the requirement of reproducing the field equations when substituted into the Euler–Lagrange equations. The Lagrangian densities are:

$$\mathcal{L} = \frac{1}{2}(\partial_\mu \phi \, \partial^\mu \phi - m^2 \phi^2) \quad \text{for the Klein–Gordon field,}$$

$$\mathcal{L} = -\frac{1}{4} F_{\mu\nu} F^{\mu\nu} \quad \text{for the electromagnetic field, and}$$

$$\mathcal{L} = \bar{\psi}(i\gamma^\mu \partial_\mu - m)\psi \quad \text{for the Dirac-field}$$

where ψ is the four-component Dirac field (column) and $\bar{\psi}$ is defined as $\bar{\psi} = \psi^* \gamma^0$, an adjoint (row) multiplied by γ^0, referred to as the Dirac adjoint, which is simply a matter of notational convenience that became a standard notation.

The canonical quantization procedure in terms of the commutators, as shown above, is rooted in the Poisson bracket formalism of Hamilton's formulation of mechanics, as discussed in Chapter 3. It leads to a successful theory of quantized fields for the Klein–Gordon and electromagnetic fields, that is, those that represent particles of spin zero and one, in fact, of all integer values of spin. The particles of half-integer spins, half, and one and half, and so on, must satisfy the Pauli exclusion principle and the fields that represent these particles, the Dirac field in particular, must be quantized not by commutators

$$[A, B] \equiv AB - BA$$

but by anticommutators

$$\{A, B\} \equiv AB + BA.$$

The choice of commutators versus anticommutators depending on whether the spin has integer or half-integer value can be compactly expressed as

$$AB - (-1)^{2s} BA$$

where s stands for either integer or half-integer. The anticommutators have no classical counterparts, that is, there are no such things as Poisson antibrackets, but nevertheless the quantization by anti-commutators is one of the fundamental requirements for quantizing fields that correspond to particles of half-integer spins. In the case of Dirac fields, the origin of the use of anticommutators can be traced back to γ^{μ} matrices that are required, by definition, to satisfy anti-commutation relations among them.

Another item to be mentioned here is concerned with the wide use of the term "second quantization." When referring to the quantization of matter fields, the term accurately describes the situation. The equations for matter fields — Klein–Gordon and Dirac fields — are actually relativistic quantum mechanical wave equations. That is the "first" quantization. The wave equations are then viewed as "classical" field equations, in an emulation of the classical electromagnetic field, and then quantized again. That is the "second" quantization. As far as the classical wave equations for electromagnetic fields are concerned, however, this is not accurate. For the electromagnetic field, the wave equation is a classical wave equation and its quantization is its "first" quantization. The term "second quantization," however, picked up a life of its own and became synonymous with quantum field theory and is widely used interchangeably with the latter. Strictly speaking, such interchangeable use of the two terms is not entirely accurate, especially where quantization of the classical electromagnetic field is concerned.

9

Particles and Fields III:
Particles as Quanta of Fields

Quantum field theory presents the third and the latest stage in the evolution of the concept of particles. This concept has evolved from that of a localized point mass in classical physics to that of wave–particle duality in quantum mechanics and, as shown in this chapter, to that of a quantum of quantized field. As classical fields are quantized, following the road map outlined in the last chapter, we will see that the concept of particles has become secondary to that of quantized fields. The quantal structure of fields, or more precisely the quantal structure of energy and momentum of fields, defines particles as discrete units of the field carrying the energy and momentum characteristics of each particle. In this sense, fields play the primary physical role and particles only the secondary role as units of discrete energy of a given field.

When the electromagnetic four-potential $A_\mu(x)$ is quantized, according to standard procedure outlined in the previous chapter, it becomes a field operator, much the same way that q's and p's, the coordinates and momenta, turn into operators in quantum mechanics. The field operator $A_\mu(x)$ consists of two parts — this is common property for all fields, whether Klein–Gordon or Dirac — called the positive frequency and negative frequency parts. The positive frequency part corresponds to raising the energy of an electromagnetic system by one unit of quantum and this corresponds to the creation operator of a photon. Likewise, the negative energy frequency part corresponds to lowering the energy of an electromagnetic system by one unit of quantum and this is the annihilation operator of a photon. The successful incorporation of photons, the zero-mass particles of light, into the fold of quantized electromagnetic four-potential was in fact the catalyst, as discussed in previous chapters, for the reinterpretation of relativistic wave equations of particles as "imitation" matter fields which started the whole ball rolling toward today's quantum field theory of particles.

The description of quantization of fields — electromagnetic, Klein–Gordon, Dirac and others — is the first-order of business for any graduate-level textbooks of quantum field theory and is usually featured in a substantial part of such textbooks. Quantization of the Dirac field alone, for example, usually takes two to three long chapters to discuss all the relevant details. We will refer to any one of the standard textbooks for full details of field quantization[1] and strive only to bring out its essential aspects in this chapter. In order to illustrate the emergence of the creation and annihilation operators of the field operators, it is very helpful first to briefly review the operator techniques involved in the case of simple harmonic oscillator problem of non-relativistic quantum mechanics.

Consider a one-dimensional simple harmonic oscillator. Denote the energy of the system by H (for Hamiltonian, which is equal to the total energy) as

$$H = \frac{p^2}{2m} + \frac{1}{2}m\omega^2 x^2$$

where m is the mass of the oscillating particle and ω is the classical frequency of oscillation. The expression above for energy can also be

---

[1]For example, such classics as *An Introduction to Relativistic Quantum Field Theory* by S. Schweber (1961), *Relativistic Quantum Fields* by J. Bjorken and S. Drell (1965), and *Introduction to Quantum Field Theory* by P. Roman (1969).

denoted as

$$H = \left(a^* a + \frac{1}{2} \right) \omega = \left(a a^* - \frac{1}{2} \right) \omega$$

where (recall that $\hbar = 1$ in the natural unit system)

$$a = \frac{1}{\sqrt{2m\omega}} (i\, p + m\omega x)$$

$$a^* = \frac{1}{\sqrt{2m\omega}} (-i\, p + m\omega x).$$

The basic commutation relation between x and p

$$[x, p] = i$$

leads to the following commutation relations,

$$[a, a] = [a^*, a^*] = 0$$

and

$$[a, a^*] = 1.$$

As is well known from elementary non-relativistic quantum mechanics, the lowest energy (the ground state) of the system is equal to $\frac{1}{2}\omega$ and the operators a^* and a, respectively, increase or decrease energy by the quantized amount equal to ω. The quantized energy in units of ω represents the quantum of the system and the operators a^* and a, respectively, raise and lower the energy by the oscillator. For this reason a^* and a, are respectively called the raising and lowering operators, for a simple harmonic oscillator.

When we quantize a field, an exactly analogous situation occurs: the field operator consists of "raising" and "lowering" operators that increase or decrease the energy of the system described by the field and it is the "quantum" of that energy that corresponds to the particle described by the field. The "raising" and "lowering" operators of the field are called creation and annihilation operators and the "quantum" of energy corresponds to a particle described by the field. Although it was the quantization of electromagnetic field that preceded that of matter fields, we will illustrate this procedure for the simplest case: Klein–Gordon field which is a scalar field (spin zero) and real (complex fields describe charged spin zero particles). The quantization of electromagnetic and Dirac fields has the added complications of having to deal with spin indices for photons (polarizations of photons) and half-integer spin particles such as electrons.

The case of Klein–Gordon field is specified by, as discussed in Chapter 8:

Field: $\phi(x)$
Field equation: $(\partial^\mu \partial_\mu + m^2)\phi(x) = 0$
Lagrangian density: $\mathcal{L} = \dfrac{1}{2}(\partial_\mu \phi \partial^\mu \phi - m^2 \phi^2)$
Momenta field: $\pi(\mathbf{x}, t) \equiv \dfrac{\partial \mathcal{L}}{\partial(\partial\phi/\partial t)} = \dfrac{\partial\phi}{\partial t}.$

The field equation, that is, the Klein–Gordon equation, allows plane-wave solutions for the field $\phi(x)$ and it can be written as

$$\phi(x) = \frac{1}{(2\pi)^{3/2}} \int b(k)e^{ikx} \, dk$$

where $kx = k^0 x^0 - \mathbf{kr}$, $dk = dk^0 d\mathbf{k}$ and $b(k)$ is the Fourier transform that specifies particular weight distribution of plane-waves with different k's. As a solution of the field equation, there is a restriction on the transform $b(k)$, however. Substituting the plane-wave solution into the field equation shows that $b(k)$ has the form

$$b(k) = \delta(k^2 - m^2)c(k)$$

in which $c(k)$ is arbitrary. The delta function simply states that as the solution of Klein–Gordon equation, the plane-wave solution must obey Einstein's energy–momentum relation, $k^2 - m^2 = 0$.

The Einstein's energy–momentum relation ($k^2 - m^2 = 0$) also places a constraint on dk, and as we shall see, this constraint is one of the basic ingredients of quantization of all fields. The integral over dk is not all over the $k^0 - \mathbf{k}$ four-dimensional space, but rather only over $d\mathbf{k}$ with k^0 restricted by the relation, (for notational convenience we switched from $(k^0)^2$ to k_0^2)

$$k_0^2 - \mathbf{k}^2 - m^2 = 0.$$

Introducing a new notation

$$\omega_k \equiv +\sqrt{\mathbf{k}^2 + m^2} \quad \text{with only the } + \text{ sign,}$$

either $k_0 = +\omega_k$ or $k_0 = -\omega_k$. Integrating out k_0, the plane-wave solutions decompose into "positive frequency" and "negative frequency" parts.[2] This decomposition, which is basic to all relativistic fields, matter fields as well as the electromagnetic field, has nothing to do with field quantization and is rooted in the quadratic nature of Einstein's energy–momentum formula. The plane-wave solutions can be written in the form:

$$\phi(x) = \int d^3\mathbf{k}(a(\mathbf{k})f_k(x) + a^*(\mathbf{k})f_k^*(x))$$

where

$$f_k(x) = \frac{1}{\sqrt{(2\pi)^3 2\omega_k}} e^{-ikx} \quad \text{and} \quad f_k^*(x) = \frac{1}{\sqrt{(2\pi)^3 2\omega_k}} e^{+ikx}.$$

The integral is over $d^3\mathbf{k}$ only and $a(\mathbf{k})$ and $a^*(\mathbf{k})$ are the respective Fourier transforms for "positive frequency" and "negative frequency" parts. Two remarks about the standard practice of notations are called for here: The star superscript (*) stands for complex conjugate in classical fields, but when they are quantized and become non-commuting operators, the notation will stand for Hermitian adjoint. After decomposition into "positive frequency" and "negative frequency" parts, the notation k_0, as in e^{-ikx}, stands as a shorthand for $+\omega_k$, that is, after k_0 is integrated out, notation $k_0 = +\omega_k$. For

---

[2]See Appendix 4 for more details.

brevity, we often write

$$\phi(x) = \phi^{(+)}(x) + \phi^{(-)}(x)$$

with

$$\phi^{(+)} = \int d^3\mathbf{k}\, a(\mathbf{k}) f_k(x) \quad \text{and} \quad \phi^{(-)}(x) = \int d^3\mathbf{k}\, a^*(\mathbf{k}) f_k^*(x).$$

We now quantize the field by imposing the canonical quantization rule, mentioned in Chapter 8, namely:

$$[\phi(\mathbf{x}, t), \phi(\mathbf{x}', t)] = [\pi(\mathbf{x}, t), \pi(\mathbf{x}', t)] = 0$$
$$[\phi(\mathbf{x}, t), \pi(\mathbf{x}', t)] = i\delta^3(\mathbf{x} - \mathbf{x}')$$

where $\pi(\mathbf{x}, t) \equiv \partial\mathcal{L}/\partial(\partial\phi/\partial t) = \partial\phi/\partial t$. These commutation rules become commutation relations among $a(\mathbf{k})$'s and $a^*(\mathbf{k})$'s, thus:

$$[a(\mathbf{k}), a(\mathbf{k}')] = [a^*(\mathbf{k}), a^*(\mathbf{k}')] = 0$$
$$[a(\mathbf{k}), a^*(\mathbf{k}')] = \delta^3(\mathbf{k} - \mathbf{k}').$$

These commutation relations are essentially identical to those for raising and lowering operators of the simple harmonic oscillator. The quantal structure of the quantized field, and resulting new interpretation of particles, is then exactly analogous to the case of raising and lowering operators for a simple harmonic oscillator.

We define the vacuum state (no-particle state), Ψ_0, to be the state with zero energy, zero momentum, zero electric charge, and so on. When we operate on this vacuum state with operator $a^*(\mathbf{k})$, the resulting state

$$\Psi_1 \equiv a^*(\mathbf{k})\Psi_0 \quad \text{(one-particle state)}$$

corresponds to that with one "quantum" of the field that has a momentum $\mathbf{k}$ and energy $\omega_k \equiv +\sqrt{\mathbf{k}^2 + m^2}$. With $E = \omega$ and $\mathbf{p} = \mathbf{k}$ ($\hbar = 1$), this quantum is none other than a relativistic particle of mass m defined by

$$E^2 - \mathbf{p}^2 = m^2.$$

A spin zero particle of mass m thus corresponds to the quantum of Klein–Gordon field and is created by the operator $a^*(\mathbf{k})$. For this reason, the operator $a^*(\mathbf{k})$ is called the creation operator. The operator

$a(\mathbf{k})$ does just the opposite,

$$a(\mathbf{k})\Psi_1 = \Psi_0,$$

and the operator $a(\mathbf{k})$ is called the annihilation operator. Repeated application $a^*(\mathbf{k})$'s leads to two, three, ..., n-particle state; likewise repeated application of $a(\mathbf{k})$'s reduces the number of particles from a given state. The quantized Klein–Gordon field operator hence contains two parts, one that creates a particle and the other that annihilates a particle: a field operator acting on the n-particle state gives both $(n+1)$- and $(n-1)$-particle states. A relativistic particle of mass m now corresponds to the quantum of the quantized field. This is the third and, so far the final, evolution in our concept of a particle.

The quantization of an electromagnetic field is virtually identical to that discussed above for the Klein–Gordon field, except that due to the polarization degrees of freedom (the spin of photons), the field $A_\mu(x)$ requires a little more care. The polarization of the electromagnetic field has only two degrees of freedom, the right-handed and left-handed circular polarizations, but the field $A_\mu(x)$ has four indices ($\mu = 0, 1, 2, 3$). This is usually taken care of by making a judicial choice allowed by gauge transformation: we choose such a gauge in which $A_0 = 0$ and $\nabla \cdot \mathbf{A} = 0$. This choice, called the radiation gauge, renders $A_\mu(x)$ to have only two independent degrees of freedom. Besides the added complications involved in the description of polarizations, the remaining procedures in the quantization of electromagnetic field are identical to that of the Klein–Gordon field and, after quantization, the electromagnetic field operator also decomposes into creation and annihilation parts:

$$A_\mu(x) = \text{annihilation operator for a single photon}$$
$$+ \text{creation operator for a single photon}$$
$$= A_\mu^{(+)}(x) + A_\mu^{(-)}(x).$$

As the quantum of electromagnetic field, the photon is a particle that has zero mass and carries energy and momentum given by $E = |\mathbf{p}| = \omega(\hbar = c = 1)$.

The quantization of Dirac field is more involved on several accounts: first, the Dirac field ψ is a four-component object, as discussed in Chapter 5, the Lagrangian density involves not only the field ψ but also its Dirac adjoint $\bar{\psi} = \psi^* \gamma^0$, and the canonical quantization must be carried out in terms of anticommutators rather than the usual commutators, as discussed in Chapter 8. When all is said and done, the Dirac field operators decompose as follows (say, for the electron):

$$\psi(x) = \psi^{(+)}(x) + \psi^{(-)}(x)$$

$\psi^{(+)}(x)$ annihilates an electron

$\psi^{(-)}(x)$ creates a positron (anti–electron)

and

$$\bar{\psi}(x) = \bar{\psi}^{(+)}(x) + \bar{\psi}^{(-)}(x)$$

$\bar{\psi}^{(+)}(x)$ annihilates a positron

$\bar{\psi}^{(-)}(x)$ creates an electron.

To sum up, when we quantize a field, it turns into a field operator that consists of creation and annihilation operators of the quantum of that field. In the case of an electromagnetic field, the classical field of the four-vector potential turns into creation and annihilation operators for the quantum of that field, the photon. In the case of matter fields, we first reinterpret the one-particle relativistic quantum mechanical wave equations as equations for classical matter fields and then carry out the quantization. Matter particles, be they spin zero scalar particles or spin half particles such as electrons, positrons, protons and neutrons, emerge as the quanta of quantized matter fields, whether they are Klein–Gordon or Dirac fields. Essentially, this is what quantum field theory of particles is all about.

10

Emulation of Light II: Interactions

The quantization of fields and the emergence of particles as quanta of quantized fields discussed in Chapter 9 represent the very essence of quantum field theory. The fields mentioned so far — Klein–Gordon, electromagnetic as well as Dirac fields — are, however, only for the non-interacting cases, that is, for free fields devoid of any interactions, the forces. The theory of free fields by itself is devoid of any physical content: there is no such thing in the real world as a free, non-interacting electron that exerts no force on an adjacent electron. The theory of free fields provides the foundation upon which one can build the framework for introducing real physics, namely, the interaction among particles.

We must now find ways to introduce interactions into the procedure of canonical quantization based on the Lagrangian and Hamiltonian formalism. The question then is what is the clue and prescription by which we can introduce interactions into the Lagrangian densities. There are very few clues. In fact, there is only one known prescription to introduce electromagnetic interactions and it comes from the Hamiltonian formalism of classical physics, as discussed in Chapter 2. Comparing the classical Hamiltonian (total energy) for a free particle with that of the particle interacting with the electromagnetic field, the recipe for introducing the electromagnetic interaction is the substitution rule (sometimes referred to as the "minimal" substitution rule)

$$p_\mu \Rightarrow p_\mu - eA_\mu.$$

Replacing p_μ by its quantum-mechanical operator $i\partial_\mu$, we have as the only known prescription for introducing electromagnetic interaction:

$$i\partial_\mu \rightarrow i\partial_\mu - eA_\mu.$$

We switched the notation for the charge from q to e. This substitution is to be made only in the Lagrangian density of free matter fields representing charged particles, but not to every differential operator that appears in a given Lagrangian density, not, for example, to differential operators in the Lagrangian density for a free electromagnetic field.

In the last chapter, we used the simple scalar Klein–Gordon field to illustrate the process of field quantization and the resulting emergence of particles as quanta of the field. To illustrate the introduction of interaction by substitution rule, we switch from Klein–Gordon to the Dirac field. All particles of matter — electrons, protons, neutrons that make up atoms, that is, all quarks and leptons (more on these later) — are spin half particles satisfying the Dirac field equations and the description of electromagnetic interactions of these particles, say, electron, requires the substitution rule to be applied to the Lagrangian density for the Dirac field.

The Lagrangian density for charged particles, say, electrons, interacting with the electromagnetic field is then given by applying the substitution rule to the Lagrangian density for the free Dirac field, and combining with the Lagrangian density for the electromagnetic field, we have

$$\mathcal{L} = \bar{\psi}(\gamma^\mu(i\partial_\mu - eA_\mu) - m)\psi - \frac{1}{4}F_{\mu\nu}F^{\mu\nu}$$

$$= \bar{\psi}(i\gamma^\mu\partial_\mu - m)\psi - \frac{1}{4}F_{\mu\nu}F^{\mu\nu} - e\bar{\psi}\gamma^\mu\psi A_\mu.$$

For brevity, we omitted the functional arguments, (x), from all fields in the above expression, i.e. $\mathcal{L} = \mathcal{L}(x)$, $\psi = \psi(x)$, $A_\mu = A_\mu(x)$, etc. The new Lagrangian describes the local interaction of electron and photon fields at the same space–time point x. Substituting this interaction Lagrangian into the Euler–Lagrange equation, we obtain the field equations for interacting fields, which as expected, are different from the equations for free Dirac and electromagnetic fields:

$$(i\gamma^\mu \partial_\mu - m)\psi(x) = e\gamma^\mu A_\mu \psi(x)$$
$$\partial_\nu F^{\mu\nu} = e\bar{\psi}(x)\gamma^\mu \psi(x).$$

We need to make several important observations here about this new interaction Lagrangian. First, the field equations for interacting fields are highly nonlinear and they are also coupled; to solve one, the other must be solved. The Dirac and electromagnetic fields, $\psi(x)$ and $A_\mu(x)$, that appear in the interaction Lagrangian, although they have the same notation, are *not* the same as the free Dirac and electromagnetic fields. Secondly, in order to proceed with the quantization of interacting fields, as illustrated in the case of free fields in the last chapter, the first thing we need are the solutions to the coupled field equations given above. We could then presumably proceed to decompose the solutions for interacting fields and perhaps even define "interacting particle creation and annihilation operators." Once we have the exact and analytical solutions for fields satisfying the coupled equations, we may have the emergence of real, physical particles as quanta of interacting fields. Quantum field theory for interacting particles would have been completely solved, and we could have moved on beyond it. Well, not exactly. Not exactly, because no one can solve the highly nonlinear coupled equations for interacting fields that result from the interacting Lagrangian density obtained by the substitution rule. Exact and analytical solutions for interacting fields have never been obtained; we ended up with the Lagrangian that we could not solve!

Just to illustrate one key point of departure from the quantization of free fields, consider the requirement that each component of the free Dirac field must also satisfy, over and beyond the Dirac equation, the Klein–Gordon equation. The requirement that $k_0^2 - \mathbf{k}^2 - m^2 = 0$, that is, $k_0 = +\omega_k$ or $k_0 = -\omega_k$ with $\omega_k \equiv +\sqrt{\mathbf{k}^2 + m^2}$, is what allowed the decomposition of the free field into creation and annihilation operators, that, in turn, led to particle interpretation. In the case of interacting fields, this is no longer possible.

At this point, the quantum field theory of interacting particles proceeded towards the only other alternative left: when so justified, treat the interaction part of the Lagrangian as a small perturbation to the free part of the Lagrangian. We write

$$\mathcal{L}(x) = \mathcal{L}_{\text{free}}(x) + \mathcal{L}_{\text{int}}(x)$$

where

$$\mathcal{L}_{\text{free}}(x) = \bar{\psi}(x)(i\gamma^\mu \partial_\mu - m)\psi(x) - \frac{1}{4}F^{\mu\nu}(x)F_{\mu\nu}(x)$$

$$\mathcal{L}_{\text{int}}(x) = -e\bar{\psi}(x)\gamma^\mu \psi(x)A_\mu(x).$$

The perturbative approach with the Lagrangian above is the basis for quantum field theory of charged particles interacting with the electromagnetic field, to wit, the quantum electrodynamics, QED. To this date, QED, with some further fine-tuning (more on this in the next chapter), remains the most successful — and so far the only truly successful — theory of interacting particles ever devised. The perturbative approach of QED is well justified by the smallness of the charge, e, renamed the coupling constant (the fine-structure constant defined as $\alpha = e^2/4\pi$ is approximately equal to $1/137$), which ensures that successive higher orders of approximation would be smaller and smaller. In the zeroth-order, then, the total Lagrangian is equal to free Lagrangian and by the same token, in the zeroth-order, the interacting fields are equal to free fields, and successive orders in the perturbation expansion in terms of the interaction Lagrangian add "corrections" to this zeroth-order approximation, generically called the radiative corrections.

The interaction Lagrangian,

$$\mathcal{L}_{\text{int}}(x) = -e\bar{\psi}(x)\gamma^\mu \psi(x)A_\mu(x),$$

is thus the centerpiece of QED. It is a compact expression that contains, interpreted in terms of the free fields, eight different terms involving various creation and annihilation operators, for each index μ, for a total of 32 terms. For each $\mu = 0$, 1, 2, 3, the expression

$$j^\mu = \bar\psi(x)\gamma^\mu\psi(x),$$

which is a four-element row matrix times a 4×4 matrix times a four-element column matrix, expands to [creation of electron + annihilation of positron] multiplied by [creation of positron + annihilation of electron]. This $j^\mu = \bar\psi(x)\gamma^\mu\psi(x)$ is then multiplied by [creation of photon + annihilation of photon], three field operators being coupled at the same space–time point x.

The success of QED, albeit by the perturbative approach, has catapulted to the above form of interaction Lagrangian to much greater significance and is more fundamental than originally perceived; it became the mantra for all other interactions among elementary particles, namely, the weak and strong nuclear forces. The weak and strong nuclear forces, as well as the electromagnetic force, are to be written in the form

$$g\bar\Psi(x)\gamma^\mu\Psi(x)B_\mu(x)$$

where g is the generic notation for coupling constants, be it electromagnetic, weak nuclear or strong nuclear force, $B_\mu(x)$ is the generic notation for the force field of each force, and the Dirac field operators for all spin half matter fields. This expression forms the basis of our understanding of all three interactions at a local point and hence, by extension, the microscopic nature of these forces — three field operators — Dirac field, Dirac adjoint field, and the force field operators — all come together at a space–time point x.

For nonelectromagnetic interactions, weak and strong nuclear forces, the adoption of the interaction Lagrangian modeled after the electromagnetic interaction Lagrangian is basically a matter of faith and can be justified only by the success of just extension. We assume the interactions to be derivable from Lagrangian density (this assumption gets some degree of justification when viewed in terms of the so-called "gauge" fields, as will be discussed in a later chapter) and to be just as local as the electromagnetic interaction. Lacking any theoretical basis, such as the substitution rule in the case of electro-magnetic interaction, the casting of non-electromagnetic interactions in the form of the interaction Lagrangian density given above corre-sponds to a grand emulation of the electromagnetic force, to wit, an emulation of light indeed.

11

Triumph and Wane

The success of quantum electrodynamics in agreeing with and predicting some of the most exact measurements is nothing less than spectacular. The quantitative agreements between calculations of QED and experimental data for such atomic phenomena as the Lamb shift, the hyperfine structure of hydrogen, and the line shape of emitted radiation in atomic transitions are truly impeccable and has helped to establish QED as the most successful theory of interacting particles. As stated previously, this is what made QED the shining example to emulate for other interactions.

To proceed from the interaction Lagrangian density

$$\mathcal{L} = \bar{\psi}(i\gamma^\mu \partial_\mu - m)\psi - \frac{1}{4}F_{\mu\nu}F^{\mu\nu} - e\bar{\psi}\gamma^\mu\psi A_\mu$$

to the results of calculations that are in remarkable agreement with observation, however, the theory had to be negotiated through some tortuous paths — calculations that yield infinities, the need to redefine some parameters that appear in the Lagrangian density, and proof that all meaningless infinities that occur can be successfully absorbed in the redefinition program. They are respectively called the ultraviolet divergences, mass and charge renormalizations, and renormalizability of QED.

With the Lagrangian density, and the resulting highly coupled field equations, that could not be solved exactly, there was only one recourse left and that was to seek approximate solutions in which the interaction term was treated as a perturbation to the free-field Lagrangian. The smallness of the coupling constant e would seem to ensure that such perturbation approach is amply justified. But when calculations were carried out order by order in the perturbation expansion in terms of the interaction Lagrangian, the results were disastrous; calculations led to results that were infinite!

The origin of infinities is believed to be an inherent property of the canonical formalism of field theory; within the Lagrangian framework, the values of a field at every space–time point x is considered as generalized coordinates and clearly there are infinite number of generalized coordinates. For example, consider a system consisting of an infinite number of non-relativistic quantum mechanical harmonic oscillators. The ground state energy of each oscillator is $1/2\ \omega$, but the total energy of the system is, of course, infinite. As a system of infinite number of generalized coordinates, appearance of infinities in calculations is actually not surprising. The appearance of infinities is called the problem of ultraviolet divergences. The way to get around this near fatal situation is in what is called the mass and charge renormalizations.

As discussed in the last chapter, the Lagrangian density breaks up into two parts:

$$\mathcal{L}(x) = \mathcal{L}_{\text{free}}(x) + \mathcal{L}_{\text{int}}(x)$$

with

$$\mathcal{L}_{\text{free}}(x) = \bar{\psi}(x)(i\gamma^\mu \partial_\mu - m)\psi(x) - \frac{1}{4}F^{\mu\nu}(x)F_{\mu\nu}(x)$$
$$\mathcal{L}_{\text{int}}(x) = -e\bar{\psi}(x)\gamma^\mu\psi(x)A_\mu(x).$$

In the perturbation approach, we imagine the interaction Lagrangian to be switched off, in the zeroth-order approximation, and are then left with the well-established free field theory. Of course, in reality this cannot be true, no more than for us to claim that we have an electron without electromagnetic interaction! Now, there are two basic parameters that enter into the total Lagrangian, the mass m and the

charge e. Within the perturbation approach, they represent the mass
and charge of a totally hypothetical electron that has no electromagnetic interaction. The mass and charge parameters in the Lagrangian
cannot be the actual, physically measured mass and charge of a real,
physical electron. They must be recalibrated so as to correspond to
the measured values of mass and charge. This need to recalibrate the
two fundamental parameters that appear in the Lagrangian density
is called renormalization, mass and charge renormalizations.

The requirements of mass and charge renormalizations, on the one
hand, and the inescapable appearance of infinities in perturbative
calculations, on the other hand, are actually quite separate issues;
they trace their origins to different sources. In practice, however, the
two become inseparably intertwined in that we utilize the procedures
to renormalize mass and charge to absorb, and thus get rid of, the
unwanted appearance of infinities in calculations. We refer to the
mass parameter that appear in the Lagrangian as the *bare* mass, of
an electron, and change its notation from m to m_0 and define the
physically observed mass, of an electron, as m. The physical mass is
then related to the bare mass by

$$m = m_0 - \delta m.$$

Both m_0 and δm are unmeasurable and unphysical quantities. The
physically measured mass of an electron, $0.5\,\mathrm{MeV}$, corresponds to
the physical mass m defined as the difference between the bare
mass and δm, sometimes called the mass counter term. In situ-
ations where no infinities appear, the mass counter term should
be, in principle, calculable from the interaction Lagrangian. It can
then be shown in the perturbation calculations that certain types
of infinities that occur can all be lumped into the mass counter
term. With the bare mass also taken to be of infinite value, the
two infinities — the infinities coming out of the perturbation calculations and the infinity of the bare mass — cancel each other
out leaving us with a finite value for the actual, physical mass of
an electron. The difference between two different infinities can certainly be finite. This process, quite fancy indeed, is called mass
renormalization.

The procedure for charge renormalization is a bit more involved than that for mass renormalization, but the methodology is the same. The physical charge is the finite quantity that results from the cancellation of two infinities — between the infinite bare charge and certain other types of infinities that appear in the calculations, that is, types other than those absorbed in mass renormalization.

The crucial test is to show that all types of infinites that occur in the perturbation calculations can be absorbed by the recalibration procedure of physical parameters, that is, the mass and charge renormalizations. Then, and only then, solutions obtained by the perturbation expansion can be accepted. This acid requirement is called the renormalizability of a theory. The two critical requirements for a quantum theory of interacting fields are thus:

(i) Perturbation expansion in terms of the interaction Lagrangian must be justified in terms of the smallness of the coupling constant.

(ii) Such expansion is proven to be renormalizable.

QED passes these two requirements with flying colors. The question now is what about the non-electromagnetic interactions.

As discussed in the last chapter, the interaction Lagrangian for the weak and strong nuclear forces was obtained simply by emulating the format for the electromagnetic interaction. Lacking any specific guide such as the substitution rule, which is deeply rooted in the Lagrangian and Hamiltonian formalism of classical physics, all we could do for these non-electromagnetic forces was to adopt the interaction form given by

$$g\bar{\Psi}(x)\gamma^{\mu}\Psi(x)B_{\mu}(x)$$

where g is the coupling constant signifying the strength of force, $\Psi(x)$ is the relevant Dirac field — proton, neutron, electron and other Dirac fields — and $B_{\mu}(x)$ is the spin one force field. We immediately run into a brick wall when it comes to the strong nuclear interaction: the coupling constant is too large for perturbation expansion in terms of the interaction Lagrangian to be considered. In the same scale as the fine structure constant $\alpha = e^2/4\pi$ of the electromagnetic interaction being equal to $1/137$, the coupling constant for the strong nuclear interaction is approximately equal to 1. The question of the perturbation expansion in terms of the coupling constant simply goes out the window for a strong nuclear force. For a weak nuclear interaction, the problem is opposite. The coupling constant for the weak nuclear force is small enough, much smaller in fact than the fine structure constant, and this in itself would ensure the validity of perturbation expansion. Rather, the problem was renormalizability. The number of infinities that occur in the perturbation calculations far exceeded the number of parameters that could absorb them by renormalization. The theory as applied to the weak interaction was simply non-renormalizable. Spectacular triumph was noted in the case of the electromagnetic interaction on the one hand, and complete failures in the case of weak and strong nuclear interactions on the other hand. In the early 1950s, this was the situation.

Quantum field theory cast in the framework of canonical quantization — often called the Lagrangian field theory — came to its mixed ending, unassailable success of QED followed by non-expandability in the case of strong nuclear force and by non-renomalizability in the case of weak nuclear force. And thus ended what might be called the first phase of quantum field theory, the era of success of the Lagrangian field theory in the domain of electromagnetic interaction with the attempt to emulate the success of QED for the case of weak and strong nuclear forces ending up in total failure.

Starting from the 1950s, interest in the Lagrangian field theory thus began to wane and the need to make a fresh start became paramount. This was the beginning of what may be called the second phase of quantum field theory. Discarding the doctrine of the canonical quantization within the Lagrangian and Hamiltonian framework, new approaches were adopted to construct an entirely new framework: building on basic sets of axioms and symmetry requirements, constructing scattering matrices for incorporating interactions that could relate to the observed results. There have been many branches of approach in this second phase, often referred to as the axiomatic quantum field theory, and they occupied a good part of two decades,

1950s and 1960s. But in the end, the axiomatic quantum field theory could not bring us any closer to analytic solutions for interacting fields. By the end of 1960s, the hope for formulating a successful quantum field theory for non-electromagnetic forces began to dim.

Beginning with the 1970s, however, a new life was injected for the Lagrangain field theory — a new perspective on how to introduce the electromagnetic interaction and a new rationale for emulating it for non-electromagnetic interactions. It is called the "local gauge field theory." Coupled with the newly-gained knowledge of what we consider to be the ultimate building blocks of matter, this new local gauge field theory would come to define what we now call the "standard model" of elementary particles. The advent of local gauge Lagrangian field theory is the latest in the development of quantum field theory and corresponds to what may be called its third phase — canonical Lagrangian, axiomatic, and now the local gauge Lagrangian field theory.

12

Leptons and Quarks

Before we begin to delve into the development of the local gauge field theory, it is good for us to take stock of the members of the Standard Model — the cast of players, if you will. We are talking about leptons, quarks, gauge bosons, and the Higgs particle.

We will start with leptons and quarks in this chapter and will deal with other particles in subsequent chapters. The case of leptons is simpler — not necessarily simple, but simpler — than those of quarks, gauge bosons (of which the photon is the prime member), and the Higgs particle and we will begin with leptons.

The name "lepton" is derived from the Greek word "leptos" meaning small and light. When the term was first coined in 1948, the only known leptons were electrons and muons (neutrinos were assumed to be massless then) and their masses are only fractions of a proton. But the discovery in the mid 1970s of the heaviest lepton to date, the tau-lepton or simply tauon, completely dispelled the notion of leptons being light. A tauon weighs some 3,500 times that of an electron and almost twice the weight of a proton, but is still a lepton!

Actually, the more accurate definition of lepton lies in the fact that all leptons have nothing whatsoever to do with the strong nuclear force. All leptons undergo the weak nuclear force and the charged leptons (electron, muon, and tauon) interact via the electro-magnetic force, but they are totally unresponsive and indifferent to the strong nuclear force. As far as leptons are concerned, the strong nuclear force does not exist. So, the more apt name for leptons might have been 'strong-force-free particles.'

The first and primary member of the lepton family is the all-familiar electron, the quantum of electric charge and the source for all things electric and magnetic, not to mention that the swirling cloud of electrons around an atomic nucleus is the 'surface' of an atom. Dis-covered in 1897, mass of an electron checks in at $0.511\,\mathrm{MeV}/c^2$ (for a discussion of mass units of choice in particle physics, see Appendix 5). The electric charge on an electron is negative 1.6×10^{-11} Coulombs, again expressed in human-sized unit of Coulomb.

All charged particles, with the sole exception of quarks, carry integer units of the magnitude of the electron's charge, denoted by $|e| = 1.6 \times 10^{-11}$ Coulombs. Charges of particles are denoted simply as -2, -1, 0, 1, or 2 and so on. The charge of an electron is thus -1. A doubly positive charge of $+2$ stands for $2\,|e|$.

Next comes the muon. Muon, denoted by the Greek letter mu, μ, was discovered in 1936 but at first it suffered misinterpretation for some time. First thought as a strongly interacting particle, it was originally named mu-meson. Over time, however, people came to realize that it was a decay byproduct of a strongly interacting particle (pi-meson) and was not a meson at all. Named muon from then on, its mass is $105.7\,\mathrm{MeV}/c^2$, some 200 times heavier than an electron. The electric charge of a muon is -1, just like the electron.

Other than the difference in their masses, the electron and muon behave exactly the same way, that is, when energetically possible due to the high mass of muon. Whatever an electron does, so does the muon. Whatever an electron does not, the muon does not. A muon is thus an exact copy of an electron, except it is about 200 times heavier. A muon is thus a heavy electron and it remained a mystery as to why it is there and what if any is the meaning of its existence. Often quoted quip by I. I. Rabi is "Who ordered that?!"

Neutrino is a particle desperately hypothesized in 1935 by Wolfgang Pauli to balance out the energy conservation law in beta decay. The beta decay occurs when a neutron decays, by the weak force, into a proton and an electron. In order to satisfy conservation laws, the decay product electron had to be accompanied by a particle that had no mass and no electric charge. This hypothetical particle was later named neutrino, denoted by the Greek letter nu, ν.

One problem with the neutrino, ν, is its inability to interact with other particles readily. It has no use for the strong force and being electrically neutral it does not respond to the electromagnetic force either. It knows of only one force, the weak force, by which it was hypothesized to be born. Its extreme "antisocial" behavior is not hard to understand, when we realize just how weak the weak force is.

In the scale where we take the strength of the familiar electro-magnetic force to be 1, the strong force is about 100 times stronger, but the weak force is about 10^{-11} times weaker, 10 trillionth of the strength of the electromagnetic force. It took a long time to detect telltale signs of neutrino, the final confirmation of it coming in 1956.

So, from 1897 till 1960, there were only three kinds of leptons, electron, muon, and neutrino, e, μ, and ν. It all changed in 1962.

Now, under the right circumstance, a neutrino can convert itself back into an electron, and with sufficient energy it can convert itself back to a muon. An energetic neutrino should be able to convert itself back into either muon or electron, and in fact into both. But a funny thing happened in 1962 when this idea was put to test. A beam of high energy neutrinos produced in high-energy accelerator along with muons were observed to convert themselves back only to muons, and never to electrons. On the other hand, a beam of low-energy neutrinos produced in beta decays along with electrons were observed never to convert to muons, only to electrons!

Somehow, the neutrinos seem to "remember their pedigree": they "remember" which leptons, electron or muon, they were born with and will convert back only to their own partners! We came to realize that there are actually two different neutrinos — an electron neutrino, ν_e, and a muon neutrino, ν_μ. Nowadays, we know that they have different mass. But other than the mass difference, just how they actually differ from each other and how they "remember" their "electron-ness" and "muon-ness" is still something of a mystery.

Thus, by 1962 we have had four leptons: a pair of electron doublet and muon doublet, each doublet consisting of negatively charged lepton accompanied by respective neutrinos.

$$\begin{pmatrix} e^- \\ \nu_e \end{pmatrix} \quad \text{and} \quad \begin{pmatrix} \mu^- \\ \nu_\mu \end{pmatrix}$$

The post-1970 parlance is to say that we have two "generations," each generation having two "flavors," the electron generation having two flavors, electron and electron neutrino, and the muon generation having two flavors, muon and muon neutrino.

This picture of four flavors of leptons would remain intact for another 14 years when in 1976, heaviest of all leptons was discovered, the tau-lepton or tauon. A tauon has a mass of $1{,}777.8\,\text{MeV}/c^2$, some 3,500 times heavier than an electron. The ratio of the masses of electron, muon, and tauon is 1: 200: 3,500! A tauon weighs almost twice that of a proton, or equal to the weight of a deuteron!

The neat picture of leptons consisting of generations, each with two flavors, was so persuasive that the discovery of tauon immediately implied the existence of its own neutrino, a tau neutrino. The tau neutrino was eventually confirmed in 2000.

In this way we have arrived at the present understanding of the lepton family, three generations, each with two flavors — (electron and electron neutrino, muon and muon neutrino, and tauon and tauon neutrino). The electron, muon, and taon all have the same electric charge -1 and three flavors of the neutrinos are all electrically neutral and although originally assumed to be massless now have different masses, however miniscule.

$$\begin{pmatrix} e^- \\ \nu_e \end{pmatrix} \quad \begin{pmatrix} \mu^- \\ \nu_\mu \end{pmatrix} \quad \begin{pmatrix} \tau^- \\ \nu_\tau \end{pmatrix}$$

Recent experiments, while unable to pin down the masses of the neutrinos directly, were able to set upper bounds for their masses: less than $2.2\,\text{eV}/c^2$, $170\,\text{keV}/c^2$, and $15.5\,\text{MeV}/c^2$ respectively for ν_e, ν_μ, and ν_τ. Note how miniscule the mass of an electron neutrino is compared to the electron. It is less than $0.0000022\,\text{MeV}/c^2$.

Another issue with the neutrino mass is the fact that the Standard Model allows only for zero-mass neutrinos. There have been many attempts to incorporate the neutrino mass into the framework of the Standard Model, with varying degrees of success, but this issue is far from resolved as of now.

The story of quarks is of much shorter history than that of leptons, but is a bit more involved. Ever since the notion of quarks erupted onto the center stage of particle physics, it grew, became very powerful and dominated the particle physics landscape.

Ever since the idea of quarks was first introduced in 1963 (papers were published in 1964, but by 1963 it was known to the particle physics community worldwide), we worked and lived with the idea of quarks for five decades now — and still counting — and we became so familiar with quarks that they became as real as McDonald's cheeseburger, making us almost forget that they have never been directly observed to date. Many believe in the dogma called the quark confinement that an isolated quark can never be directly observed. Any and all properties of quarks are thus based on indirect inferences.

In the pre-quark days before 1963, the group of particles that constituted elementary particles — the basic building blocks of all known matter in the Universe — were divided into two distinct camps, a group of heavier particles called hadrons and a group of relatively lighter ones called leptons. The premier members of hadrons are protons, neutrons and pions (pi-mesons). A proton, for example, is about 1,874 times more massive than electron, the premier member of leptons.

The names "hadrons" and "leptons" originate from Greek words meaning "strong" and "small," respectively, although this distinction becomes meaningless as the heaviest lepton, the tauon, turns out to be about twice as massive as proton. What really separates hadrons from leptons is more dynamical in nature than the gaps in their masses: hadrons interact via the strong nuclear force whereas leptons have nothing to do with the strong nuclear force. All particles, both hadrons and leptons interact via the weak nuclear force and electrically charged hadrons and leptons via the electromagnetic force.

Careful study of the systematics of hadrons led Murray Gell-Mann and, independently, George Zweig to propose in 1963 that hadrons may be, after all, composites of still more elementary objects. Gell-Mann named them quarks while Zweig called them aces, and we all know which name became a household word.

The weird name, quarks, actually goes well with some weird properties of them: they would have to have hitherto-unheard-of fractional values of electric charges. The up quark would have to have +2/3 in units of |e| while the down and strange quarks −1/3 |e|. Note that the difference is still 1 unit, $+2/3 - (-1/3) = 1$.

Originally, three types — three flavors in modern parlance — of quarks were proposed, named up (u), down (d) and strange (s). All mesons are to be made up of a quark and an antiquark and baryons, as the extended family of proton and neutron is called, are to be made up of three quarks. A proton is considered to be a bound system of up-up-down $(2/3 + 2/3 - 1/3 = 1)$ while a neutron is of up-down-down $(2/3 - 1/3 - 1/3 = 0)$.

This picture of a triplet of quarks, u, d, and s, was not unlike the early triplet picture of leptons, e, μ, and ν. The three early leptons would become four, with the recognition of two flavors of neutrinos, an electron neutrino, ν_e, and a muon neutrino, ν_μ, and the four would group themselves into two generations of two flavors each.

Well, the same groupings would happen for quarks in 1974. In November of that year, a bombshell of announcements came forth from the East and West coasts of the United States. A new particle was discovered simultaneously at the Brookhaven laboratory in New York and the Stanford Linear Accelerator lab in California that defied conventional interpretation in terms of u, d, and s quarks. The particle was a meson all right, supposedly made up of a quark and an antiquark, but its properties were consistent with a new type of quarks beyond the three original ones. After much analyses, it became clear that the new particle, named ψ at Stanford and J at Brookhaven, was a bound system of a new quark and its antiquark. The new quark is named 'charm,' denoted by c, and it had to have the electric charge of +2/3 just like the up quark. The charm quark immediately filled up the picture of two generations of quarks, each generation with two flavors, just like the leptons.

$$\begin{pmatrix} u^{+2/3} \\ d^{-1/3} \end{pmatrix} \quad \begin{pmatrix} c^{+2/3} \\ s^{-1/3} \end{pmatrix}$$

When you compare this picture of two generations of quarks with the three-generations of leptons, the missing third generation of quarks becomes all too compelling.

$$\begin{pmatrix} e^- \\ \nu_e \end{pmatrix} \quad \begin{pmatrix} \mu^- \\ \nu_\mu \end{pmatrix} \quad \begin{pmatrix} \tau^- \\ \nu_\tau \end{pmatrix}$$

The search for the possible third generation of quarks was on! In another three years, in 1977, a team at the Fermi laboratory near Chicago would discover a new meson, named Upsilon that was consistent with being a bound system of yet another new species of quark and its antiquark. The new flavor of quark was named 'bottom' quark, denoted by b, and it was consistent with carrying the electric charge of $-1/3$, just like the down and strange quarks.

Now the search for the partner to the bottom quark, named top quark, with electric charge of $+2/3$ became the all-consuming focus of particle physics. As the anticipated top quark would not be found readily, the search became frustratingly futile. Many began to doubt the neat picture of three generations with two flavors each for quarks. It was to take 19 long years for teams at the Fermi laboratory to finally claim the victory of having found the trace of top quark in 1996. Again, by "finding the top quark" what is meant is finding a meson that is consistent with being a composite of a top quark and an antitop quark.

And now finally we arrived at the three generations picture of quarks and leptons. The six flavors of quarks and six flavors of leptons, neatly divided into three generations, represent our current understanding of what constitutes all known matter in the Universe.

$$\begin{pmatrix} u^{+2/3} \\ d^{-1/3} \end{pmatrix} \quad \begin{pmatrix} c^{+2/3} \\ s^{-1/3} \end{pmatrix} \quad \begin{pmatrix} t^{+2/3} \\ b^{-1/3} \end{pmatrix}$$

$$\begin{pmatrix} e^{-1} \\ \nu_e^0 \end{pmatrix} \quad \begin{pmatrix} \mu^{-1} \\ \nu_\mu^0 \end{pmatrix} \quad \begin{pmatrix} \tau^{-1} \\ \nu_\tau^0 \end{pmatrix}$$

The idea of quarks is not without issues, however, no matter how much we become familiar with it and in fact how much we need it to understand the workings of elementary particles.

First and foremost is the fact that no single isolated quark has ever been observed. We infer all their properties, their mass and spin, indirectly from hadrons that they are the constituents of. We sweep this frustrating fact under the rug called the quark confinement which states that an isolated quark can never be observed. The dogma of quark confinement, however plausible, is not yet an established theory, but a doctrine that harbors a little bit of hope and prayer.

The force that binds quarks into hadrons — three quarks for baryons and quark–antiquarks for mesons is an entirely new force. It is not the same strong force that binds protons and neutrons into atomic nuclei. Confusingly this new interquark force is also called the strong force, but is not the same strong force among protons and neutrons. Names are the same, but the new one is among the quarks and the old one is among the hadrons. When a proton and a neutron form a deuteron, it is the result of the new strong forces among three quarks inside the proton and three quarks inside the neutron. It resembles the "molecular forces" among the constituents of atoms that make up a molecule. The definition of the strong force has completely changed.

In terms of quarks and leptons, we now need the theoretical framework to study their interactions via all three forces, strong, weak, and electromagnetic. This was to be done within the framework of the Lagrangian quantum field theory and this brings us to the discourse on what is called the gauge field theory, the edifice of the Standard Model.

13

What is Gauge Field Theory?

As we mentioned in Chapter 11, the heydays of quantum electrodynamics was over by the early 1950s and in the next two decades, the
1950s and 1960s, the canonical Lagrangian field theory was rarely
spoken of. The 50s and 60s were primarily occupied by the search for
patterns of symmetries in the world of elementary particles — resulting in the introduction of quarks as mentioned in Chapter 12 — and
the pursuit of quantum field theory was carried out by those inves-
tigating the formal framework of the theory, generally called the
axiomatic field theory, starting from scratch seeking new ways to
deal with weak and strong nuclear forces.

During this period that may be called the second phase of
quantum field theory, the Lagrangian field theory was almost completely sidelined and the emphasis was on the formal and analytic properties of scattering matrix, the so-called S-matrix theories
and the axiomatic approaches to field theory. These new axiomatic
approaches, however, did not bring solutions to quantum field theories any closer than the Lagrangian field theories.

Beginning in the 1970s, there surged a powerful revival of the
Lagrangian field theory that continues to this day. This is what is
called the (Lagrangian) gauge field theory, and it starts — yes, once

again — from quantum electrodynamics, the photons! The gauge field theory represents the third and current phase in the development of quantum field theory.

It turns out that the Lagrangian density for QED,

$$L = \bar{\psi}(i\gamma^\mu\partial_\mu - m)\psi - \frac{1}{4}F_{\mu\nu}F^{\mu\nu} - e\bar{\psi}\gamma^\mu\psi A_\mu$$

obtained by the substitution rule and splendidly successful in describing the electromagnetic interaction between charged particles and photons, harbored yet another gift to give us.

First, it is invariant under two trivial transformations: the basic invariance of the bilinears of the type $\bar{\psi}\ldots\psi$ under simple phase transformation $\psi \rightarrow \psi e^{-i\alpha}$ where α is a real constant and, secondly, the invariance owing to the classical gauge invariance of electromagnetic field under $A_\mu \rightarrow A_\mu + \partial_\mu\Lambda$ where Λ is an arbitrary function. Note that this electromagnetic gauge invariance is valid as long as the photon mass is zero. If the photon should have had mass m, there would have to be another term in the Lagrangian $m^2 A_\mu A^\mu$ which is clearly not invariant under the gauge transformation $A_\mu \rightarrow A_\mu + \partial_\mu\Lambda$.

The set of all such transformations, $\psi \rightarrow \psi e^{-i\alpha}$, phase change with a real constant, constitute a unitary group of dimension one, a trivial group denoted as U(1) and we refer to it as the global phase transformation. We say that the QED Lagrangian is invariant under global phase transformation. It is a "big" name for something so trivial, but the idea here is to set the language straight and distinguish this trivial case from more complicated cases yet to come when phase transformations are local, that is, dependent on x, rather than global.

A few words on terminology might be in order here. Phase transformation, whether global or local, are nowadays also called "gauge" transformation. To the extent that the original definition of gauge transformation refers to the electromagnetic field, as discussed in Chapter 6, this may be a little confusing. There is a good rationale to extend the definition of gauge transformation to include the local phase transformation and this will be explained below. Until then, we will stick with the use of the name phase transformation (which is actually what it is).

Let us now consider phase transformations that are local, that is, the phase α is a function of x. Since the fields at each x are considered as independent variables in the scheme of canonical quantization formalism, it is not unreasonable to consider different phase transformations at different space–time points x. The question now is whether or not the QED Lagrangian is invariant under such local phase transformation. It is immediately clear by observation that the QED Lagrangian is not invariant under local phase transformation; all terms in the Lagrangian except one are trivially invariant, but the "kinetic energy" term involving the differential operator is not. We have

$$\bar{\psi}e^{i\alpha(x)}(i\gamma^{\mu}\partial_{\mu})e^{-i\alpha(x)}\psi = \bar{\psi}i\gamma^{\mu}\partial_{\mu}\psi + \bar{\psi}\gamma^{\mu}\psi\partial_{\mu}\alpha(x)$$

and the QED Lagrangian picks up an extra term $\bar{\psi}\gamma^{\mu}\psi\partial_{\mu}\alpha(x)$.

At this point the gauge transformation of the electromagnetic potential $A_{\mu}(x)$ swings into action. As discussed in Chapter 6, $A_{\mu}(x)$ is determined only up to four-divergence of an arbitrary function $\Lambda(\text{x})$, that is,

$$A^{\mu} + \partial^{\mu}\Lambda = \left(\phi + \frac{\partial\Lambda}{\partial t}, \mathbf{A}\text{-}\nabla\Lambda\right)$$

which is the original gauge transformation of electromagnetism.

If we now choose — this is the most important aspect of the local gauge field theory — the arbitrary function $\Lambda(x)$ to be equal to the local phase transformation of the Dirac field divided by the electromagnetic coupling constant e, that is,

$$\Lambda(x) = \frac{\alpha(x)}{e},$$

the interaction term of the Lagrangian yields another extra term that exactly cancels out the unwanted term,

$$-e\bar{\psi}\gamma^{\mu}\psi A_{\mu} \rightarrow -e\bar{\psi}\gamma^{\mu}\psi\left(A_{\mu} + \frac{1}{e}\partial_{\mu}\alpha\right) = -e\bar{\psi}\gamma^{\mu}\psi A_{\mu} - \bar{\psi}\gamma^{\mu}\psi\partial_{\mu}\alpha.$$

The interplay between the local phase transformation on the Dirac field and the matching choice of the electromagnetic gauge transformation "constructively conspires" to render the QED Lagrangian invariant under local phase transformations.

In sum, the QED Lagrangian

$$L = \bar{\psi}(i\gamma^\mu \partial_\mu - m)\psi - \frac{1}{4}F_{\mu\nu}F^{\mu\nu} - e\bar{\psi}\gamma^\mu\psi A_\mu$$

is invariant under the local phase transformation

$$\psi \rightarrow \psi e^{-i\alpha(x)}$$

provided the gauge transformation of A_μ is chosen to be

$$A_\mu \rightarrow A_\mu + \frac{1}{e}\partial_\mu \alpha(x)$$

With this choice, the local *phase* transformation on the Dirac field becomes interwoven with the electromagnetic gauge transformation and changes its name to local *gauge* transformation.

The freedom of gauge transformation of the electromagnetic potential thus plays an indispensable role without which the invariance under local gauge transformation cannot be upheld. There is one more absolutely crucial role played by the gauge invariance. The proof of renormalizability of QED discussed in Chapter 11 crucially depends on the gauge invariance of the QED Lagrangian. As we mentioned above, the gauge invariance requires the electromagnetic field to correspond to *massless* spin one particles, to wit, photons. The roles played by gauge transformation of A_μ within the framework of QED are absolutely indispensable.

The invariance of QED Lagrangian under the local gauge transformation is now to be elevated to the lofty status of a new general principle of quantum field theory, which can perhaps be extended to interactions other than electromagnetic, namely, the weak and strong interactions. To this end, we can now state the new principle, christened the *gauge principle*, as follows: From the way in which the freedom of gauge transformation of the electromagnetic field $A_\mu(x)$ plays the crucial role, we can define a generic field, say $B_\mu(x)$, with just such property and call it the gauge field. A gauge field is defined as a four-vector field with the freedom of gauge transformation, and it corresponds to massless particles of spin one.

The gauge principle requires that the free Dirac Langrangian $L = \bar{\psi}(i\gamma^\mu \partial_\mu - m)\psi$ be invariant under the local gauge transformation $\psi \to \psi e^{-i\alpha(x)}$. The invariance is upheld when we invoke a gauge field $B_\mu(x)$ such that

(i) $\partial_\mu \to \partial_\mu + igB_\mu(x)$

(ii) $B_\mu(x) \to B_\mu(x) + \dfrac{1}{g}\partial_\mu \alpha(x)$

where g stands for the coupling constant of a particular interaction, that is, the strength of a particular force. This new gauge principle then leads to a unique interaction term of the form $g\bar{\psi}\gamma^\mu\psi B_\mu$. What the gauge principle does is that it reproduces the substitution rule as a consequence of the invariance of free Dirac Lagrangian under the local gauge transformation, thereby bypassing the classical Hamiltonian formalism for charged particles in an electromagnetic field.

To be sure, this new invariance of the QED Lagrangian does neither enhance nor reduce any aspect of the success of QED. As we mentioned before, QED based on the substitution rule does just fine without invoking the invariance under the local gauge transformation. What the local gauge invariance does is to provide further insight into the nature of the QED Lagrangian and most importantly it provides an entirely new window of opportunity to formulate and advance theories for the weak and the strong forces.

Once again — for the third time in fact — the electromagnetic interaction provides a path of emulation for other interactions to follow. The most critical element in this new approach is the idea of gauge fields and this is how the gauge field theory, or more fully gauge quantum field theory of interacting particles, was born.

In the case of QED, there is one and only one gauge field, namely the electromagnetic field, that is, one and only type of photons. Photon is in a class by itself and does not come in a multiplet of other varieties. In the language of representation of a group, the photon is a singlet. Local gauge transformation involves pure numbers that are functions of x. The set of all such local gauge transformations form a one-dimensional trivial group U(1) which is by definition commutative. Non-commutativity will involve matrices rather than pure numbers. There is another name for being commutative called Abelian, non-commutative being non-Abelian. We refer to the U(1) local gauge transformation also as Abelian U(1) transformation.

In the case of weak and strong interactions, the situation becomes much more complex. The symmetries involved dictate the gauge fields to come in multiplets, compared to a single gauge field, the electromagnetic field A_μ, in the case of the electromagnetic force. In the case of weak interactions, we need three gauge fields and for strong interactions, we need eight gauge fields. Applying the gauge principle to the weak and the strong interactions is definitely more complicated.

14

The Weak Gauge Fields

In quantum physics, be it quantum mechanics or quantum field theory, we deal with SU(N) transformations. It is with a good reason.

We are familiar with rotations in the real space, that is, where the coordinates are all real numbers, such as x, y and z. All such rotations form a group called SO(N), rotations in N-dimensional real space. Why O? Rotations in real space are mathematically called orthogonal transformations.

In quantum physics, we always deal with complex numbers. All wave functions and quantum fields are complex functions and we need to consider "rotations" in complex space. Why U? "Rotations" in complex space are mathematically called unitary transformations. All "rotations" in complex space form a group denoted as SU(N), "rotations" in N-dimensional complex space. The phase transformation $\psi \rightarrow \psi e^{-i\alpha(x)}$ is an SU(1) transformation. In one-dimensional case, we drop S and just call it U(1).

As we discussed in Chapter 12, quarks and leptons fall into the pattern of three generations, each with two-component column, two flavors, and they define a two-dimensional complex space. For brevity, we will call them q^+ for (u, c, and t) and q$^-$ for (d, s, and b) for quarks, and for leptons l^- for (e, μ, and τ) and ν for the three neutrinos. The three generations can be compactly expressed as

$$\begin{pmatrix} q^+ \\ q^- \end{pmatrix} \quad \text{and} \quad \begin{pmatrix} l^- \\ \nu \end{pmatrix}$$

The appropriate symmetry for quarks and leptons is then an SU(2) symmetry of the two flavors.

Now there are historically at least two other SU(2) symmetries before the new one for quarks and leptons, and a few words are in order so that we can properly differentiate the different SU(2)s.

The oldest and the most familiar SU(2) symmetry is that of the particle spin. All quarks and leptons are spin $1/2$ (in units of $\hbar$) particles and their spin up and spin down states form the two dimensional column,

$$\begin{pmatrix} \uparrow \\ \downarrow \end{pmatrix}$$

All SU(2)$_{spin}$ rotations are generated by the three 2×2 matrices mentioned in Chapter 5, namely

$$\sigma_1 = \begin{pmatrix} 0 & 1 \\ 1 & 0 \end{pmatrix}, \quad \sigma_2 = \begin{pmatrix} 0 & -i \\ i & 0 \end{pmatrix}, \quad \text{and} \quad \sigma_3 = \begin{pmatrix} 1 & 0 \\ 0 & -1 \end{pmatrix}.$$

The next SU(2) symmetry is that of the isotopic spin of proton and neutron. Other than the fact that there is a very small mass difference between the two and that proton is positively charged while neutron is uncharged, as far as the strong nuclear force is concerned, proton and neutron can be viewed as two different states of a single particle, referred simply as nucleon. The idea of the isotopic spin, or simply isospin, is to treat proton and neutron as the isospin up and down states of a nucleon, just as the spin up and down states of an electron. The SU(2)$_{isospin}$ is generated by the identical three 2×2 matrices, except we change the symbol from σ to τ since the notation σ has long been associated with the mechanical spin. So, to repeat, we have

$$\tau_1 = \begin{pmatrix} 0 & 1 \\ 1 & 0 \end{pmatrix}, \quad \tau_2 = \begin{pmatrix} 0 & -i \\ i & 0 \end{pmatrix}, \quad \text{and} \quad \tau_3 = \begin{pmatrix} 1 & 0 \\ 0 & -1 \end{pmatrix}.$$

Now, the $SU(2)_{\text{isospin}}$ is defined only with respect to the strong nuclear force and is not defined for leptons that have nothing to do with the strong force.

For the two-component doublets of quarks and leptons, we need to define a new $SU(2)$ that is different from $SU(2)_{\text{isospin}}$. The new $SU(2)$ for both quarks and leptons is called $SU(2)_{\text{weak isospin}}$ and the old $SU(2)_{\text{isospin}}$ must now be differentiated from the new and renamed $SU(2)_{\text{strong isospin}}$. The $SU(2)$ we must deal with for both quarks and leptons in the Standard Model is the $SU(2)_{\text{weak isospin}}$. With this understood, we now call the new weak isospin $SU(2)$ simply as $SU(2)$. Had enough of $SU(2)$?

When we now apply the new local gauge principle to formulate a theory for weak nuclear force, we must go through a few layers of complications. The first complication we run into has to do with increased number of the local phase functions from one to three. In the $U(1)$ case of QED, a single phase function $\alpha(x)$ was all we needed. With the $SU(2)$ symmetry of the weak force case, we need to have the local phase functions to be a scalar with respect to $SU(2)$ as well and the required expression is

$$\tau_k \alpha_k(x) = \tau_1 \alpha_1(x) + \tau_2 \alpha_2(x) + \tau_3 \alpha_3(x).$$

The compact expression $\tau_k \alpha_k$ is deceptively simple but explicitly spelled out in terms of 2×2 matrices, it looks like this:

$$\tau_k \alpha_k(x) = \begin{pmatrix} \alpha_3(x) & \alpha_1(x) - i\alpha_2(x) \\ \alpha_1(x) + i\alpha_2(x) & -\alpha_3(x) \end{pmatrix}$$

The simple $U(1)$ local gauge transformation for the electromagnetic force in terms of a single function $\alpha(x)$,

$$\exp(-i\alpha(x)),$$

is now replaced, in the case of the weak force, by $SU(2)$ local gauge transformation in terms of three phase functions,

$$\exp\left[-i\begin{pmatrix} \alpha_3(x) & \alpha_1(x) - i\alpha_2(x) \\ \alpha_1(x) + i\alpha_2(x) & -\alpha_3(x) \end{pmatrix}\right],$$

and the local gauge transformation $\exp(-i\tau_k\alpha_k(x))$ is now applied to two-component field of quarks and leptons.

The more serious complication — and this is what renders the weak gauge field theory to be non-Abelian gauge field theory — arises from the fact that the 2×2 matrices τ_i of SU(2) are non-commutative. They satisfy the commutation relation

$$[\tau_i, \tau_j] = i\varepsilon_{ijk}\tau_k$$

where ε_{ijk} is the usual structure constant of the SU(2) algebra. One can readily see how this non-commutativity relations lead to complications when we impose the local gauge transformation. The non-commutativity raises its head right away: the kinetic energy term in the free Lagrangian for the Dirac fields of quarks and leptons becomes

$$\bar{\psi}e^{i\tau_j\alpha_j}i\gamma^\mu\partial_\mu e^{-i\tau_k\alpha_k}\psi = \bar{\psi}i\gamma^\mu\partial_\mu\psi + \bar{\psi}e^{i\tau_j\alpha_j}[-\gamma^\mu\tau_k\partial_\mu\alpha_k]e^{-i\tau_k\alpha_k}\psi.$$

In the second term, the factor $e^{-i\tau_k\alpha_k}$ must be commuted through τ_k in the middle before the two phase factors can be collapsed, and this will bring in the structure constant term into gauge transformation of the weak gauge fields.

The relatively "simple" recipe in the case of QED, that is,

$$\partial_\mu \to \partial_\mu + igB_\mu(x)$$

$$B_\mu(x) \to B_\mu(x) + \frac{1}{g}\partial_\mu\alpha(x)$$

is replaced by

$$\partial_\mu \to \partial_\mu + ig\tau^k W_\mu^k(x)$$

$$W_\mu^k(x) \to W_\mu^k(x) + \frac{1}{g}\partial_\mu\alpha^k(x) + \varepsilon^{klm}\alpha^l(x)W_\mu^m$$

where $W_\mu^k(x)\,(k=1,2,3)$ are the three weak gauge fields and the term involving ε^{klm} is new one arising out of the non-Abelian properties and is needed to cancel out the extra term resulting from commuting the 2×2 matrices.

So in going from U(1) of QED to SU(2) of weak interaction, things did get a little more complicated, but nevertheless the recipe of the local gauge invariance we discovered from QED can be successfully applied to the case of the weak interaction. We have single gauge field $A_\mu(x)$, the photon field, for QED and for the weak interaction we have three gauge fields $W_\mu^k(x)$. The task of formulating the gauge field theory for the weak interaction is now complete, right? Well, not quite. Not so fast.

Remember the key word 'renormalizability' in QED that was absolutely essential for the theory to be completed? All previous attempts to formulate Lagrangian field theory for the weak interaction failed to the wayside because they were not renormalizable. The new gauge field theory of the weak interaction remains renormalizable if and only if the gauge fields $W_\mu^k(x)$ are totally massless. All three must correspond to zero mass particles.

Now, to be sure, the non-Abelian gauge theory we have just outlined above is not something entirely new, not something developed after the developments of quarks and leptons. Rather, it was originally considered back in 1954 by C. N. Yang and R. L. Mills. The Yang–Mills theory, as it is called, actually predates the idea of quarks by about ten years. The gauge fields of the original Yang–Mills theory had to be massless and since the particles that mediate the strong and the weak forces had to be rather massive, the adaptation of Yang–Mills theory for these forces was out of the question and the theory remained an interesting but unrealistic idea for almost two decades.

There is an inverse proportionality relation between the range of a force and the mass of the particle that is mediating the force (Appendix 6). The shorter the range, heavier the mass and vice versa. Photon has no mass and it is consistent with the range of the electromagnetic force being infinite. The ranges for the weak and the strong forces are, respectively, 10^{-18} and 10^{-15} meters and the mediating particles had to be rather massive.

The three weak gauge fields $W_\mu^k(x)$ also have to be massless, in order to uphold the requirement of renormalizability and it appears to suffer the same fate, interesting but unrealistic idea. The actual weak bosons that mediate the weak force are in fact quite massive, to put it mildly. Of the three weak bosons, W^+, W^-, and the neutral Z bosons, the idea of charged W bosons has been around for sometime

but without precise prediction for their masses due to lack of satis-
factory field theory for the weak force, while the idea of neutral Z
boson is relatively of recent vintage. It wasn't until the early 1980s
that the mass for these weak bosons were pinpointed; W bosons were
determined to have mass of 80 GeV in 1983 and Z boson to have the
mass of 91 GeV in 1984. That is GeV, not MeV. 91 GeV corresponds
to an atomic nuclei of atomic number around 92!

So, here we have a situation, a beautiful local gauge field theory
based on SU(2) of the weak interaction on one hand and, on the
other hand, the reality of very heavy (compared to other elementary
particles) weak bosons. We seem to be between a rock and a hard
place!

What saved the day was the idea of what is called the "spontaneous symmetry breaking" and its specific application called the
"Higgs mechanism" that will pull the magic of endowing mass to
what starts out as massless fields while upholding the sacred duty of
being renormalizable. It is a fancy mathematical magic, to which we
now turn to in the next chapter.

15

The Higgs Mechanism and the Electroweak Gauge Fields

Spontaneous symmetry breaking refers to the situation where the system in a symmetrical state contains solutions that are in an asymmetrical state. It is relevant in the study of condensed matter physics such as the phase transition, ferromagnetism, and superconductivity. It is a spontaneous process, that is, without any external influence, wherein the Lagangian is symmetric, but the lowest energy state of that Lagrangian is not symmetric.

The importance of this concept to the particle physics was first noted by Yoichiro Nambu in 1960 and its application to the specific models of gauge symmetry was advanced in 1964 by three independent groups, Englert and Brout, Higgs, and Guralnik–Hagen–Kibble.

The spontaneous symmetry breaking of gauge symmetry provides an ingenious method to bestow mass to gauge fields that was massless in the Lagrangian to start with. Since the SU(2) gauge fields must be massless to ensure the renormalizability of the theory, but the particles that these massless gauge fields represent have large mass, as we discussed in the last chapter, this spontaneous symmetry breaking mechanism plays absolutely critical role in formulating the gauge theory of weak interaction.

A few words here are in order as to how the name of this method came to be called the Higgs mechanism. The mechanism should be correctly called Englert–Brout–Higgs–Guralnik–Hagen–Kibble mechanism. It is clearly too cumbersome. Even if we abbreviate it as EBHGHK mechanism, it is still too long! In 1970, in his research into the depth of this mechanism, Benjamin Lee decided to shorten the name to simply Higgs mechanism. The shortened name caught on and, as they say, the rest is history and now it is Higgs mechanism, Higgs field, Higgs particle, Higgs this and Higgs that! Higgs himself somewhat regretted the name, saying he wished it was called something else!

The workings of the Higgs mechanism, or more precisely, the spontaneous symmetry breaking mechanism of the SU(2) non-Abelian gauge symmetry is not the easiest thing in the world to explain and understand. At one time, a science minister of the United Kingdom posted a handsome award to anyone who could explain the concept to lay people.

The concept can, however, be described simply in terms of high-school algebra.

Consider two simple functions $y_1 = x^2$ and $y_2 = x^4$. They are both even functions with left-right symmetry. At $|x| = 1$, $y_1 = y_2 = 1$. For $|x| > 1$ y_2 dominates over y_1 but for $|x| < 1$ the role reverses and y_1 dominates over y_2. It is shown in the next figure with the dashed line being y_1 and the solid line being y_2.

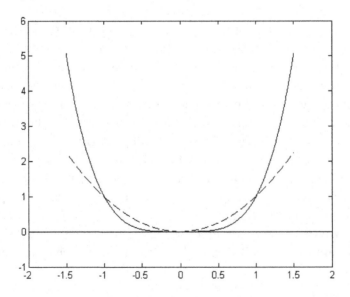

Now let us consider a specific combination $y_3 = -y_1 + y_2 = -x^2 + x^4$. The combination looks like the next graph.

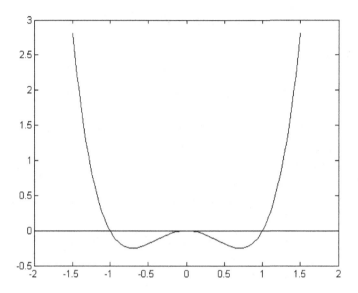

We refer to such shape as wine-bottle shape, for obvious reason. As an example of purely algebraic functions, the shape has no particular meaning other than its curious shape.

But now suppose the shape corresponds to some potential function for a quantum mechanical system. The ground state for either x^2 or x^4 near $x = 0$ is no longer the ground state. There are states with lower energy.

$\frac{dy_3}{dx} = -x(-2 + 4x^2) = 0$ has solutions not only at $x = 0$ but also at

$$x = \pm\frac{1}{\sqrt{2}} \equiv \pm a.$$

The new lowest energy state — new ground state — located at $x = +a$, or $x = -a$, has now clearly lost the left-right symmetry of the potential function defined by y_3. *This is the very essence of the spontaneous symmetry breaking: A symmetric system can yield solutions that break the symmetry of the system.*

We can now shift the coordinate x by $x = z + a$ and expanding x^2 and x^4, the function y_3 becomes, in terms of z,

$$y_3 = z^4 + 4az^3 + (6a^2 - 1)z^2 + 2a(2a^2 - 1)z + a^2(a^2 - 1),$$

showing the odd functions z^3 and z.

We can change the wine-bottle shape by considering the more general case of $-px^2 + qx^4$ with suitable values for p and q. The example given above is a trivially simple case, but it does serve to bring out the meaning of the spontaneous symmetry breaking.

The Higgs mechanism, that is, the spontaneous symmetry breaking mechanism of the SU(2) non-Abelian gauge symmetry, works in exactly the same way as the simple example described above.

A new scalar field (spin zero) field $\varphi(x)$ is introduced into the Lagrangian in addition to the Dirac field for quarks and leptons, and the three SU(2) gauge fields $W_\mu^k(x)$ such that the scalar field enters the Lagrangian in the form

$$p\varphi(x)^2 + q\varphi(x)^4.$$

This produces the wine-bottle shape of potential (for positive q) and yields solutions that break the gauge symmetry and necessitates "shifting" the field $\varphi(x)$ by

$$\varphi(x) = \omega(x) - \sqrt{|p|/2q}$$

The factor $\sqrt{|p|/2q}$ reduces to $1/\sqrt{2}$ as in the case of the simple example above when $|p| = q = 1$. The "shifted" scalar function $\omega(x)$ is what has come to be called the Higgs field.

This "shifting" to the new scalar field $\omega(x)$, the Higgs field, provides the mechanism for endowing the weak gauge fields $W_\mu^k(x)$ that were originally massless to gain a term in the Lagrangian that can be interpreted as mass terms for them, as a consequence of the symmetry breaking, spontaneously. Now, how this comes about can be better explained when applied to the cases of quarks and leptons. To be sure, quarks and leptons are not required to be massless, not by renormalizability or any other requirements. Whereas the renormalizability of the theory does require gauge fields to be massless, nothing in the theory demands quarks and leptons to be massless. But how a field corresponding to a massless quark or lepton acquires a mass term in the Lagrangian by the Higgs mechanism best illustrates how the mechanism works.

The free Lagrangian for field ψ, be it a quark field or a lepton field, is

$$\bar{\psi}i\gamma^\mu\partial_\mu\psi - m\bar{\psi}\psi \quad \text{for } m \neq 0$$

and

$$\bar{\psi}i\gamma^\mu\partial_\mu\psi \quad \text{for } m = 0.$$

We start out with massless field but introduce a new interaction (a new force!) in which the quark or lepton field interact with the scalar field $\varphi(x)$ with the coupling strength g. Thus we have as the Lagrangian for this case as

$$\text{L} = \bar{\psi}i\gamma^\mu\partial_\mu\psi + g\varphi(x)\bar{\psi}\psi.$$

The scalar field needs to be "shifted" as above

$$\varphi(x) = \omega(x) - \sqrt{|p|/2q}$$

And the Lagrangian becomes

$$\text{L} = \bar{\psi}i\gamma^\mu\partial_\mu\psi + g\omega(x)\psi\psi - g\sqrt{|p|/2q}\,\psi\psi.$$

Now if we identify $g\sqrt{|p|/2q}$ as m, the Lagrangian above describes quark or lepton field with non-zero mass interacting with the Higgs field $\omega(x)$ with the strength g.

A zero mass field ψ acquires a mass $m = g\sqrt{|p|/2q}$ by the Higgs mechanism. The mass, however, depends on three new parameters, g, p, and q and the term $g\omega(x)\bar{\psi}\psi$ describes the interaction of quarks and leptons with the Higgs field. The mechanism of how the gauge fields, $W_\mu^k(x)$, acquires their mass is a little more complicated than the case for quarks and leptons, but the basic idea is the same.

The fact that the Higgs mechanism can be invoked to create a term in the Lagrangian that looks just like a mass term is an unexpected bonus. But what a bonus! It ignited a firestorm of hype that all the mass of known matter in the Universe was 'created' by the

Higgs field $\omega(x)$ — quarks, leptons, atoms, molecules, rocks, stars, galaxies, everything! The quantum of Higgs field, the Higgs particle, was referred to as the "God particle" in a book by Leon Lederman way before the Higgs particle was 'discovered' but it lit up a firestorm of hype in media: "Here is finally our deepest understanding of what mass is and where it comes from" or "Without the Higgs particle, the Universe would have had no mass at all." The Higgs particle began to acquire somewhat divine meaning! As we have shown above, all that the Higgs mechanism does is to create a term in the Lagrangian that looks like a mass term, and does it without ruining the renormalizability. It is definitely an ingenious method, but divine it is not.

At this point we have achieved a successful SU(2) gauge field theory for weak interaction. The theory is renormalizable and at the same time the weak gauge fields $W_\mu^k(x)$ are endowed mass by the Higgs mechanism. Together with the good old U(1) gauge theory of electromagnetism, we seem to have succeeded in applying the local gauge symmetry to two of the three basic forces, the electromagnetic and weak forces. But the success of the Standard Model goes beyond the separate formulation of gauge field theories for the two forces; it provides a framework in which these two forces are 'unified,' that is, they are two different aspects of a single force, dubbed the 'electroweak' force. This is done as follows: We start out with four gauge field, one for U(1) and three for SU(2), $B_\mu(x)$ and $W_\mu^k(x)$. The real photon field $A_\mu(x)$, the charged weak boson field $W_\mu^\pm(x)$ and the neutral weak boson field $Z_\mu(x)$ are defined in terms of the four primary fields as

$$W_\mu^\pm(x) = \frac{1}{\sqrt{2}}\left(W_\mu^1(x) \pm iW_\mu^2(x)\right)$$

$$A_\mu(x) = \cos\theta_w B_\mu(x) + \sin\theta_w W_\mu^3(x)$$

$$Z_\mu(x) = -\sin\theta_w B_\mu(x) + \cos\theta_w W_\mu^3(x)$$

where the mixing angle θ_w is called the Weinberg angle. The Higgs mechanism does the trick at this point: it leaves the photon to remain massless, but endows mass to W's and Z particles. As we mentioned before, the mass of charged W bosons was accurately determined to be 80 GeV in 1983 and the neutral Z boson was discovered in 1984 with the mass of 91 GeV, all consistent with the extreme short range of the weak force.

This new U(1) × SU(2) gauge field theory in which the electromagnetic and weak forces are interwined is the unified theory of the 'electroweak' force. It is one of the two sectors of the Standard Model, the other being the sector for the strong force that deals only with quarks and not with leptons.

16

The Higgs Particle

The Higgs particle that we hear so much about nowadays is the quantum of the Higgs field $\omega(x)$ and it has spin zero since the Higgs field is a scalar field, as opposed to gauge fields whose quanta are spin one particles. Nothing in the theory predicts the value of the mass of this particle, however. The latest result from the Large Hadron Collider puts its mass at around 125 GeV.

The Standard Model consists of quarks, leptons, gauge bosons and the Higgs particle. The gauge bosons are the photons for the electromagnetic force, the weak bosons $W^{\pm}$ and Z for the weak force, and the gluons for the strong force (more on gluons in the next chapter). The Higgs particle interacts with all members of this set, except photons and gluons, for the simple reason that the Higgs particle is the agent for the spontaneous symmetry breaking mechanism and the QED as well as the theory for the strong force do not need such mechanism.

Over and above their interaction with quarks, leptons and the $W^{\pm}$ and Z bosons, the Higgs particles have a unique characteristic for interacting among themselves. This self-interaction comes about as the result of its being the agent of the spontaneous symmetry breaking.

As we discussed in the last chapter, the lowest energy state of the potential defined by

$$p\varphi(x)^2 + q\varphi(x)^4$$

requires that we shift the field as

$$\varphi(x) = \omega(x) - \sqrt{|p|/2q}.$$

This yields terms in the Lagrangian that are cubic and quartic in Higgs field, that is, terms containing $\omega(x)^3$ as well as $\omega(x)^4$ and this gives rise to Feynman diagrams of self-interaction vertices of three and four Higgs particles at a point.

A similar self-interaction in the electroweak gauge field theory — and this is a signature property of all non-Abelian gauge theory — comes about as a result of the non-commutativity of SU(2) algebra. As we mentioned in Chapter 14, the gauge transformation of the weak gauge field acquires a term that contains the structure constant of the symmetry group SU(2), thus,

$$\partial_\mu \to \partial_\mu + ig\tau^k W_\mu^k(x)$$

$$W_\mu^k(x) \to W_\mu^k(x) + \frac{1}{g}\partial_\mu\alpha^k(x) + \varepsilon^{klm}\alpha^l(x)W_\mu^m$$

The same thing happens with the definition of the tensor fields of the SU(2) gauge fields.

The antisymmetric tensor field $F_{\mu\nu}$ for the electromagnetic field A_μ is defined as

$$F_{\mu\nu} = \partial_\mu A_\nu - \partial_\nu A_\mu,$$

but the corresponding gauge field tensor must be defined by

$$F_{\mu\nu}^i = \partial_\mu W_\nu^i - \partial_\nu W_\mu^i - g\varepsilon^{ijk}W_\mu^j W_\nu^k.$$

The antisymmetric field tensor of the weak gauge fields picks up an extra term containing the structure constant of SU(2). In general, the appearance of the terms containing the structure constant is the generic property of all non-Abelian SU(N) group and this is what, among others, sets the Yang–Mills theory apart from the Abelian gauge theory of QED. In the case of the trivial U(1) group, there are no commutation relations and hence no terms involving the structure constant.

The new term in the gauge tensor field has far-reaching consequences. In contradistinction to the case of QED wherein photons do not carry electric charges and do not interact with other photons, the gauge particles of the gauge fields in the Yang–Mills theory interact with each other, and they do so with the coupling constant g, the same coupling constant with which they interact with quarks, leptons, and $W^{\pm}$, and Z bosons.

This self-interaction of the gauge fields among themselves is a striking departure from QED and is the signature hallmark of all non-Abelian gauge field theories. The imposition of the local gauge principle originally extracted from QED on the non-Abelian symmetries resulted in an entirely new class of interactions, the self-interactions of the gauge fields among themselves.

The discovery of a new scalar boson announced by two independent groups at the Large Hadron Collider (LHC) in 2012 is certainly consistent with what Higgs boson would be. But so far the evidences collected are those involving the interactions of the new particle with the quarks, leptons and $W^{\pm}$, Z bosons. While they are impressively consistent with the particle being the long sought-after Higgs particle, it would be compelling if we find evidences for the signature properties of the Higgs particle, namely, its interaction with each other, as the terms $\omega^3(x)$ and $\omega^4(x)$ demand. Until then it is not unsafe to say that the new particle is a new scalar particle of some sort that interacts with quarks, leptons and $W^{\pm}$, Z bosons in a manner that is consistent with what we would expect from a Higgs particle. With its large mass of 125 GeV, it would take energies higher than what is now available at the LHC to hope to see these self-interactions and we would have to wait till the LHC attains its highest energy level sometimes in 2016 and beyond.

In October of 2013, the Nobel Prize Committee awarded the 2013 Nobel Prize in Physics to François Englert and Peter W. Higgs for their theoretical work pointing to the existence of the Higgs particle.

17

Evolution of the Strong Force

For over nine decades since its early days in the 1930s, our knowledge of the atomic nucleus — its constituents and their inner workings — has evolved through many phases to reach its current state. The new force among the new constituents — the nuclear force — would also go through several stages of refinements and redefinitions.

The discovery of neutron (n) in 1932 immediately raised the question of just what force was holding protons (p) and neutrons together within the atomic nucleus. Since the neutron is electrically neutral, the new force is clearly not the electromagnetic force. Studies soon showed that the forces between neutron and neutron, proton and neutron, and proton and proton are the same force. This new force was then called, appropriately, the nuclear force. It was an entirely new force of nature over and beyond the electromagnetic and gravitational forces.

As far as the new nuclear force was concerned, protons and neutrons were essentially the same particles. Other than the fact that proton is positively charged while neutron is uncharged and tiny differences in their mass (938 MeV vs. 937 MeV for n and p, respectively), protons and neutrons behaved exactly in the same manner as far as the new nuclear force was concerned. A new generic name was coined for protons and neutrons called nucleon, denoted by N, and this had led to the idea of isotopic spin, the $SU(2)_{\text{isospin}}$ that we mentioned in Chapter 14.

Just as an electron has two spin states, spin up and spin down, a nucleon would have two isospin states, isospin up (proton) and isospin down (neutron). Just as the mechanical spin defines the $SU(2)_{\text{spin}}$ symmetry, the isospin defines the $SU(2)_{\text{isospin}}$ symmetry.

$$N = \begin{pmatrix} p \\ n \end{pmatrix}$$

Modeled after the electromagnetic force as the exchange of photons between the charged particles, the nuclear force was viewed as the exchange of pions, denoted by π, between nucleons. There would have to be three pions, π^+, π^- and π^0, positively charged, negatively charged, and neutral pions for isospin raising, lowering, and labeling functions of $SU(2)_{\text{isospin}}$.

$$\pi = \begin{pmatrix} \pi^+ \\ \pi^0 \\ \pi^- \end{pmatrix}$$

The nucleon and pion correspond to a doublet and a triplet column of $SU(2)_{\text{isospin}}$.

This picture of the nuclear force based on the π–N system with the $SU(2)_{\text{isospin}}$ symmetry was very effective for about two decades. It helped establish firmly the basis for the study of the nuclear force — its brute strength but very short effective range. It was a very nice and relatively simple picture that, however, would not last long.

Due to the extremely short effective distance of the nuclear force — in the order of a few fermis, one fermi being 10^{-15} meters — it is necessary to probe the internuclear distances with very high energy and this gave birth to the era of particle accelerators with ever higher energies. Today's highest-energy particle accelerator is the 20-billion dollar Large Hadron Collider (LHC) located in Switzerland which is designed to accelerate proton beams up to $14\,\text{TeV}$, that is, trillion-electronvolts.

The need for higher and higher energies for probing beams of particles — whether protons or electrons — created a 'catch-22' situation. The Einstein's mass-energy formula, $E = mc^2$ comes into play. The higher the energy, the more new particles with higher masses are created. What started as a quest to find laws of nuclear force had turned into a manufacturing plant for newer and heavier particles!

Starting in the 1950s, and continuing to this day, the high-energy particle accelerators around the world kept discovering newer and heavier members of the extended family of nucleons and pions. By the end of 1960, we accumulated no less than 200 such newer particles until we ran out of Greek letters to name them — Λ, Σ, Ξ, Δ, Σ^*, Ξ^*, Ω, K, η, ψ, Υ and many, many more. There were times when we all carried a small pocketbook containing information on some 200 new particles — we used to call those pocketbooks 'Sears catalogue.'

These new particles, discovered at a dizzying speed at the laboratories, would very quickly decay down to lower mass particles in quick sequences until they reach the familiar pions and nucleons. They would decay almost as soon as they were created and hardly qualified to be even called particles. So we called them resonances.

It soon became clear that there were at least two distinct modes of their decay — one that involved very fast decays, even by the standard of particle physics, and the other decays with much slower rate. It soon became clear that the two modes of decay were the results of two different forces in action. The fast decays were the work of a "strong" nuclear force and the slower ones the work of a "weak" nuclear force. The two forces were found to be entirely different forces and we came to realize that there is not one nuclear force, but rather two distinct forces operating within the realm of atomic nucleus.

New forces, the strong and the weak, would replace the old name of nuclear force. We now have the "strong nuclear force" and the "weak nuclear force." Eventually the names would be shortened to just the strong force and the weak force. The weak force is the one we already met; it is the force we discussed as the force mediated by $W^\pm$ and Z bosons. The strong force is the original nuclear force between pions and nucleons.

The study of the new particles under the strong and the weak forces now branched out as a new field of physics named the high-energy physics or simply particle physics. The old nuclear physics was split into two distinct fields of physics, the high-energy and low-energy nuclear physics. The low-energy nuclear physics would continue to study the structure and dynamics of atomic nuclei and eventually be renamed simply as the nuclear physics — *nuclear physics* for the study of atomic nuclei and *particle physics* for the study of elementary particles and new resonances.

As more and more higher-mass particles and resonances were discovered, it became clearer and clearer that hundreds of these particles could not possibly be all 'elementary particles.' After all, the periodic table lists only 92 naturally-occurring elements. Thus began in the mid 1950s all-out efforts to classify and understand systematics of the bewildering situation.

People began to sort these particles by their physical properties — their mass, spin, decay patterns and others — and collect them by common similarities and group them into separate folders, to see if we can spot some telling patterns.

In 1960, Murray Gell-Mann and, independently, Yuval Ne'eman recognized striking patterns — nucleons and six other resonances shared many common properties and pions and five other resonances did likewise. Each formed an octet of eight similar particles. Gell-Mann called it the 'eightfold way.' Later, more 'extended families' of nucleons appeared to form a ten-member multiplet, that is, a decuplet.

The octet and decuplet correspond to multiplet structures of a symmetry group $SU(3)$, the rotation group in 3-dimensional complex space. We have gone from $SU(2)$'s to one dimension higher to $SU(3)$. It is a symmetry much more approximate than $SU(2)$ symmetry of nucleons, the mass differences among the members of an octet being much larger than the mass difference between proton and neutron, but the grouping them as members of an $SU(3)$ octet made great sense. The $SU(2)_{\text{strong isospin}}$ symmetry for nucleon is now extended to $SU(3)_{\text{unitary spin}}$ for an octet of particles, the word 'unitary' coming from the U of SU(3) meaning rotation in complex space. The SU(3)$_{\text{unitary spin}}$ stands for the symmetry of the strong nuclear force. (The SU(2)$_{\text{weak isospin}}$ symmetry for the weak nuclear force was actually established several years after SU(3)$_{\text{unitary spin}}$.)

The discovery of the SU(3)$_{\text{unitary spin}}$ for the strong nuclear force then led to the possibility that all these new particle families, including nucleons and pions, could not be really as elementary as we thought but perhaps were actually composite structures of something more fundamental and elementary objects that define the symmetry of an SU(3), a triplet of particles that can make up the octets and decuplets.

In 1963, Murray Gell-Mann who advocated the 'eightfold way' and, independently, George Zweig came up with just such a scheme, an SU(3) triplet that can build up octets and decuplets as composite structures. Gell-Mann named them 'quarks' while Zweig called them 'aces' and we all know which choice became the household word. And this is how, very briefly, the idea of quarks was born!

As we already discussed in Chapter 12, three quarks were originally proposed, up, down and strange, u, d and s. The question that immediately arose was then what is the force that binds these quarks into nucleons, pions, and related higher mass particles and resonances. This new force is not the same force that binds protons and neutrons to form atomic nuclei. Curiously, these new force between quarks is also called the strong force! The force we now call strong force is *not* the strong force that we used to call! The name stayed the same but its definition has completely changed.

The force between nucleons that we used to call strong force is now relegated to a lesser position as "molecular" type of force. A nucleon–nucleon force is now the result of the this "new" strong force among six quarks, three quarks in one nucleon and three quarks in the other nucleon. Whereas the internucleon force is mediated by exchange of pions, the new interquark force is to be viewed as being mediated by exchange of new particles, which would come to be called gluons.

18

The History of Color SU(3) Symmetry

The gauge field theory for the strong force, the interquark force, now called quantum chromodynamics, traces its origin to 1965. That year Moo-Young Han (author of this book) and Yoichiro Nambu spelled out basic framework for such a gauge field theory after discovering hitherto hidden new SU(3) symmetry for quarks, at one level deeper than the $SU(3)_{\text{unitary spin}}$. A new re-interpretation of this Han–Nambu SU(3) symmetry and formal launching of the gauge field theory were put forward by Harald Fritzsch and Murray Gell-Mann in 1971. The new SU(3) was named the color SU(3), that is, $SU(3)_{\text{color}}$, and the particles mediating the interquark force was named gluons.

Let's briefly trace out the history of how $SU(3)_{\text{color}}$ came into being.

Since the introduction of the quark model in 1963, five decades ago, the scope of its success is truly impressive. The breath and depth with which the quark model provides the basis for our understanding of dynamics of the strong force among the hadrons are absolutely indisputable. That is not to say, however, that the quark model is without a few disturbing shortcomings.

From the very outset, the quark model had to struggle with two problems. The fact that quarks are fractionally charged, $+2/3$ and $-1/3$, unlike all other known particles was difficult to be accepted at first and even more seriously, in the case of some members of the resonances forming an $SU(3)_{unitary\ spin}$ decuplet were bound states of three identical quarks, which was in direct violation of the Pauli's exclusion principle, one of the most sacrosanct principles in quantum physics. The apparent violation of the exclusion principle was so serious that it almost killed the quark model.

In 1965, Han and Nambu proposed a theory called "Three-Triplet Model with Double SU(3) Symmetry" in which the two problems mentioned above could be dealt with in one go — to kill two birds with one stone.

First was to render the electric charges of quarks to be more "normal," that is, integrally charged like all other known particles. Each quark would come in three, up_1, up_2, and up_3, and likewise for down and strange quarks (this was before the discovery of the charm quark). The three up quarks would have electric charges 1, 1, and 0. The average is then $2/3$, the original charge assignment for the up quarks. Three down quarks, as well as three strange quarks, would have electric charges 0, 0, and -1. The average corresponds to $-1/3$, the original charge assignment for down and strange quarks. The tripling of the number of quarks led to discovery of an entirely new three-fold symmetry. Han and Nambu called them (lack of imagination!) $SU(3)''$ and the usual $SU(3)_{unitary\ spin}$ was called $SU(3)'$.

The assignment of integer charges to quarks, however, gradually fell out of favor. For one thing no quarks, whether integrally charged or fractionally charged, have ever been detected directly for all this time.

Another reason is that in order to make the quark charges integer, one has to connect the electric charges with the properties of the new $SU(3)''$ and this made the theory much more complicated and cumbersome.

The discovery of the new $SU(3)''$, however, solved the Pauli exclusion principle beautifully. What at first appeared to be bound states of three identical quarks, *uuu*, *ddd* and *sss*, were actually bound states of three different states of each quark, identical with respect to $SU(3)'$, that is, $SU(3)_{\text{unitary spin}}$, but different with respect to $SU(3)''$.

$$q = \begin{pmatrix} q_1 \\ q_2 \\ q_3 \end{pmatrix} \quad \text{where } q = u, d, \text{ and } s$$

The discovery of $SU(3)''$ tripled the number of quarks, from 3 to 9 and when we include later members of quarks, c, t and b, from 6 to 18.

Based on this newly discovered $SU(3)''$ symmetry, Han and Nambu proposed a gauge field theory by introducing eight gauge vector (spin 1) fields that correspond to an octet in $SU(3)''$ and a new interquark force mediated by these gauge bosons to be named the superstrong force. Much later, the eight gauge bosons would be named gluons and the superstrong force named simply strong force, as we discussed in the last chapter.

Tripling the number of quarks by threefold certainly seem to make things more complicated without direct experimental evidences for such a need. Its major achievement was that it solved the problem of conflict with the exclusion principle. The idea of the existence of another layer of symmetry in the form of $SU(3)''$ definitely was a great move forward but still it lacked hard evidence to support it. And the idea did not catch the attention of physics community for several years.

In his book "Higgs" published in 2012, Jim Baggott wrote "Nobody took much notice. Han and Nambu had taken a big step toward the ultimate solution, but the world was not yet ready."

Beginning in the early 1970s, direct experimental evidences began to be discovered that directly supported tripling the number of quarks.

First such evidence was found in the decay of neutral pions into two photons. A neutral pion (π^0) would split into a pair of quark and antiquark, each of which then emit photons. The theoretical calculation of this pion decay would agree with the experimentally observed value only if the number of quarks is tripled.

Another supporting evidence came in from the electron–positron colliding beam experiments carried out at the Stanford Linear Accelerator Center (SLAC). The process by which an electron and a positron annihilate each other and results in production of a quark and an antiquark, which then produce showers of known hadrons (nucleons, mesons, and their extended families of resonances) also required tripling the number of quarks and antiquarks for the theory and experiment to agree.

These two experiments established beyond any doubt the existence of the SU(3)″ discovered by Han and Nambu.

Prompted by these developments, Fritzsch and Gell-Mann made a striking and bold proposal that provided a new interpretation of this SU(3)″ symmetry:

1. Quarks would carry an entirely new kind of charges, in addition to the electric charges.
2. The new charges would have three values. They named these new charges the 'color' charges. The name 'color' has nothing to do with its usual meaning; it is just a three-valued label called 'red,' 'green,' and 'blue.'
3. The SU(3)″ symmetry would be an exact symmetry, not an approximate symmetry as all previous SU(2)'s and SU(3)'s. It means that the three quarks of different color have exactly same mass. The SU(3)″ was renamed the color SU(3) symmetry, to wit, $SU(3)_{color}$.
4. The octet of gauge vector particles was renamed 'gluons.'
5. The $SU(3)_{color}$ was to be separated from any connection to the electric charges and the quarks would carry the originally assigned values of $+2/3$ and $-1/3$.

Based on these new interpretations of SU(3)″ and the introduction of the color charges as new property for quarks, Fritzsch and Gell-Mann spelled out the details of the gauge field theory of the strong force, along the same path as the gauge field theory of the weak force that we described in previous chapters.

And this is how we came to meet the $SU(3)_{color}$ of today and the resulting theory known as the quantum chromodynamics, QCD for short.

To be sure, there was another entirely different approach to solve the apparent violation of the exclusion principle. Oscar Greenberg suggested in 1965 that perhaps we allow three identical quarks to be in the same state provided they obey new parastatistics, that is, the quarks would be 'para-particles' different from all other known particles that obey the Pauli exclusion principle. This approach is mathematically equivalent to SU(3) but it does so at the cost of provoking new rules for quarks.

19

Quantum Chromodynamics, QCD

The re-interpretation in 1971 by Fritzsch and Gell-Mann of the $SU(3)''$ symmetry of Han and Nambu was as drastic as it was foretelling. The main points of their idea are, to repeat what was stated in the last chapter:

1. Quarks carry an entirely new kind of charges that are to the strong force what the electric charges are to the electromagnetic force. This new charges are tri-valued and they define the $SU(3)''$ symmetry. The charges are named color — red, green, and blue and the $SU(3)''$ would be renamed $SU(3)_{color}$. Each quark would come in three color states,

$$q = \begin{pmatrix} red \\ green \\ blue \end{pmatrix}, \quad q = u, d, s, c, t, \text{ and } b$$

2. The $SU(3)_{color}$ was to be an exact symmetry, that is, three different color states of a given quark have exactly the same mass.
3. The $SU(3)_{color}$ was not to be related to other forces, neither electromagnetic nor weak; it is an exclusive symmetry of the strong force only.
4. The octet of gauge vector particles were renamed 'gluons.'

To counterdistinguish color states of a quark from the six species of quarks, three generations of doublets, the species of quarks are sometimes referred to as 'flavors.' In this parlance, six flavors of quarks, each with three colors make up 18 quarks altogether.

Based on these propositions, Fritzsch and Gell-Mann articulated on the gauge field theory of $SU(3)_{\text{color}}$ along the same path as the gauge field theory of $SU(2)_{\text{weak isospin}}$, as we have already discussed, with one big difference. Since the gluons are taken to be zero-mass particles, the renormalizability is guaranteed and there is no need to invoke the Higgs mechanism. The resulting theory, with unbroken exact symmetry, is what is called the quantum chromodynamics, QCD for short.

The mathematics of an $SU(3)$ is understandably more complicated than that of an $SU(2)$.

Whereas $SU(2)$ is generated by three 2×2 matrices,

$$\tau_1 = \begin{pmatrix} 0 & 1 \\ 1 & 0 \end{pmatrix}, \quad \tau_2 = \begin{pmatrix} 0 & -i \\ i & 0 \end{pmatrix}, \quad \text{and} \quad \tau_3 = \begin{pmatrix} 1 & 0 \\ 0 & -1 \end{pmatrix}$$

$SU(3)$ is generated by eight 3×3 matrices λ^k with $k = 1, 2, 3, 4, 5, 6, 7,$ and 8, called the Gell-Mann matrices, and the local phase involves eight local phase functions $\alpha^k(x)$.

$$\lambda^k \alpha^k(x) = \begin{pmatrix} \alpha^3 + \alpha^8/\sqrt{3} & \alpha^1 - i\alpha^2 & \alpha^4 - i\alpha^5 \\ \alpha^1 + i\alpha^2 & -\alpha^3 + \alpha^8/\sqrt{3} & \alpha^6 - i\alpha^7 \\ \alpha^4 + i\alpha^5 & \alpha^6 + i\alpha^7 & -2\alpha^8/\sqrt{3} \end{pmatrix}$$

The mathematics of QCD is quite complicated and no one knows how to solve the field equations of it.

Not being able to solve field equations analytically is not something new only to QCD, however. No one can solve QED, the most successful quantum field theory to date, to obtain analytic solutions of its field equations, its success coming from the perturbation approximations. The same holds for the electroweak gauge field theory, its success so far coming from lowest order calculations. In both cases, QED and the electroweak theory, calculations based on perturbation approximation is well justified by the fact that the coupling constants, that is, the strength of the forces, are small enough for expansion in terms of them to be valid.

In the case of QCD for the strong force, the usual technique of perturbation approximation does not hold in general due to the brute strength of the strong force. So the theory of QCD that was beautiful to look at remained more or less semi-active until an unexpected property of non-Abelian gauge field theory was discovered in 1973 by David Politzer and, independently by David Gross and Frank Wilczek.

In the process of calculating the self-interaction of gluons, they discovered a remarkable property of non-Abelian gauge field theory for quarks: the strength of the force, expressed as the coupling constant, is function of the interaction energy in such a way that as the distance between quarks gets shorter and shorter the coupling constant becomes smaller and smaller and at extreme short distances quarks behave as if they were free non-interacting particles. This unique property is called the 'asymptotic freedom!'

This new property is exactly opposite to what we are familiar with. Both the electromagnetic and gravitational forces obey the inverse-square law and at shorter distances, the strength of the forces skyrockets ever stronger. Asymptotic freedom is counterintuitive to us, but it does allow for the limited application of the idea of perturbation expansion in terms of coupling constants for those reactions that can occur at very short distances between quarks. The QCD at this point divides into two camps, perturbative QCD and non-perturbative QCD.

Shorter distance means higher energy interaction between quarks and thus the perturbative QCD is limited to those interactions at very high energy, but most of the issues involving quarks such as the dogma of confinement are in the domain of non-perturbative QCD and it is here that we do not have clear solutions of QCD simply because the theory is so untenably complicated to handle.

One issue of the non-perturbative QCD that is not yet fully settled is the question of the confinement. The confinement involves two related topics, more general 'color confinement' and specific case of it called the 'quark confinement.' They are not the same, the latter being one specific case of the overall dogma of the former.

Quark confinement states that at large distances the force between quarks become so strong that one cannot pull out a single isolated quark from the known particles that are made of three quarks (nucleon and its extended families) and a quark–antiquark bound sustems (pions and its extended families). It is just the opposite of the asymptotic freedom and however plausible it may be, this form of confinement, the quark confinement, is not yet rigorously proved.

The hypothesis of color confinement is much broader than the quark confinement. It decries that only the color-neutral multiplets of SU(3) can be observed and any multiplets that are not color-neutral are forbidden to exist in an observable form.

The color-neutral multiplets are singlets with respect to $SU(3)_{color}$. All known and observed particles correspond to $SU(3)_{color}$ singlets, but quarks themselves are triplets in color SU(3) and gluons are octets with respect to color and by this decree of color confinement they are denied actual observability. Another example of bound systems that are not color-neutral are diquarks, the bound states of two quarks, that are denied observability simply because they are not color-neutral. Three-quark system (qqq) is observable — protons and neutrons and their extended family of resonances — but the two-quark system (qq) is not. That is the color confinement. And however plausible it may be, the issue of the color confinement that subsumes the quark confinement is not yet satisfactorily settled.

QCD represents our best shot at understanding the nature of the strong force, that is, the interquark force, but the theory is just too complicated to unravel its secret and obtain a successful description of the strong force at all energy levels. The perturbative QCD applies only to high energy and short distance behaviors and the more general non-pertubative QCD presents us with formidable challenge in attempting to extract information. Such phenomenon as what we call hadronization is still not yet clearly understood. Hadronization refers to the production of jets of known particles from quarks and gluons. But the QCD is a very powerful and elegant theory and hopefully it will soon begin to yield results that will shed more light on the nature of the strong force.

Appendix 1: The Natural Unit System

Relativistic quantum field theory is an intricate infusion of the special theory of relativity characterized by the constant c, the speed of light, and quantum theory characterized by the constant $\hbar$, the Planck's constant h divided by 2π. It is convenient to use as the system of units consisting of these two constants plus an arbitrary unit for length, say, meters. Such a system is called the natural unit system. In terms of the standard MKS system of units, they have the values:

$$c = 3 \times 10^8 \, \text{m/sec}$$
$$\hbar = 1.06 \times 10^{-34} \, \text{Joule} \cdot \text{sec or m}^2 \, \text{kg/sec}$$
$$\hbar/c = 0.35 \times 10^{-42} \, \text{kg} \cdot \text{m}.$$

In the natural unit system, mass and time are expressed in terms of $m^{-1}c^{-1}\hbar$ and mc^{-1}, respectively, where m stands for meters.

It is also customary in relativistic quantum field theory to set $c = \hbar = 1$. Thus all physical quantities are expressed as powers of a length unit, say, meters. With this choice of dimensions, energy, momentum and mass become inverse lengths. The natural unit system with $c = \hbar = 1$ provides convenience to theoretical expressions since the two constants appear in virtually all formulas in relativistic quantum field theory. When a result of a theoretical calculation needs to be compared with experimental data, however, one has to reinstate the values of c and $\hbar$. In the world of elementary particles, masses as well as energies and momenta are usually expressed in MeV or GeV (mega-electron-volt or giga-electron-volt) and the length in terms of fm (fermi) which is equal to 10^{-15} meters. For example,

$$\hbar c \approx 197 \, \text{MeVfm}$$
$$\frac{e^2}{4\pi} \approx 1.44 \, \text{MeVfm}.$$

Appendix 2: Notation

The coordinates in a three-dimensional space are denoted by $\mathbf{r} = (x, y, z)$ or $\mathbf{x} = (x^1, x^2, x^3)$. Latin indices i, j, k, l take on space values 1, 2, 3. The coordinates of an event in four-dimensional space–time are denoted by the contravariant four-vector (c and $\hbar$ are set to be equal to 1 in the natural unit system, Appendix 1)

$$x^\mu = (x^0, x^1, x^2, x^3) = (t, x, y, z).$$

The coordinates in four-dimensional space–time are often denoted, for brevity, simply by $x = (x^0, x^1, x^2, x^3)$ without any Greek indices, especially when used as arguments for functions, as in $\phi(x)$. Greek indices $\mu, \nu, \lambda, \sigma$ take on the space–time values 0, 1, 2, 3. The summation convention, according to which repeated indices are summed, is used unless otherwise specified.

The covariant four-vector x_μ is obtained by changing the sign of the space components:

$$x_\mu = (x_0, x_1, x_2, x_3) = (t, -x, -y, -z) = g_{\mu\nu}x^\nu$$

with

$$g_{\mu\nu} = \begin{pmatrix} 1 & 0 & 0 & 0 \\ 0 & -1 & 0 & 0 \\ 0 & 0 & -1 & 0 \\ 0 & 0 & 0 & -1 \end{pmatrix}$$

The contravariant and covariant derivatives are similarly defined:

$$\frac{\partial}{\partial x^{\mu}} = \left(\frac{\partial}{\partial t}, \nabla \right) = \partial_{\mu}$$

and

$$\frac{\partial}{\partial x_{\mu}} = \left(\frac{\partial}{\partial t}, -\nabla \right) = \partial^{\mu}.$$

The momentum vectors and the electromagnetic four-potential are defined by

$$p^{\mu} = (E, \mathbf{p})$$

and

$$A^{\mu} = (\phi, \mathbf{A}),$$

respectively.

Appendix 3:
Velocity-Dependent Potential

The velocity-dependent potential within the Lagrangian formalism for the case of charged particles in an electromagnetic field has far-reaching consequences in the development of quantum field theory. It is from this velocity-dependent potential that the substitution rule for the electromagnetic interaction is derived. As such it is the very foundation for the development of quantum electrodynamics, QED. The principle of local gauge invariance of the QED Lagrangian is an abstraction based on the substitution rule. The non-Abelian gauge field theories come from applying this principle of local gauge invariance to the cases of weak and strong nuclear interactions. The genesis of the non-Abelian gauge field theories, therefore, can be traced all the way back to the discovery of the velocity-dependent potential in the 19th century. Despite such paramount importance, the subject is treated often peripherally in textbooks on classical mechanics. Here, we will briefly sketch out how the velocity-dependent potential came about.

The electric and magnetic fields in vacuum can be expressed in the form

$$\mathbf{B} = \nabla \times \mathbf{A}$$

and

$$\mathbf{E} = -\nabla\phi - \frac{\partial \mathbf{A}}{\partial t}$$

where $\mathbf{A}$ is the vector potential and ϕ the scalar potential ($c = 1$ in the natural unit system). The Lorentz force formula, $\mathbf{F} = q(\mathbf{E} + \mathbf{v} \times \mathbf{B})$, can then be written as

$$\mathbf{F} = q\left(-\nabla\phi - \frac{\partial \mathbf{A}}{\partial t} + \mathbf{v} \times (\nabla \times \mathbf{A})\right).$$

Using the identity

$$\mathbf{v} \times (\nabla \times \mathbf{A}) = \nabla(\mathbf{v} \cdot \mathbf{A}) - (\mathbf{v} \cdot \nabla)\mathbf{A},$$

the Lorentz force equation can be further rewritten as

$$\mathbf{F} = q\left[-\nabla\phi - \frac{\partial \mathbf{A}}{\partial t} + \nabla(\mathbf{v} \cdot \mathbf{A}) - (\mathbf{v} \cdot \nabla)\mathbf{A}\right].$$

Combining the gradient terms, we have

$$\mathbf{F} = q\left[-\nabla(\phi - \mathbf{v} \cdot \mathbf{A}) - \left(\frac{\partial \mathbf{A}}{\partial t} + (\mathbf{v} \cdot \nabla)\mathbf{A}\right)\right].$$

The vector potential $\mathbf{A}$ is a function of x, y, z as well as of time t and the total derivative of $\mathbf{A}$ with respect to time is

$$\frac{d\mathbf{A}}{dt} = \frac{\partial \mathbf{A}}{\partial t} + (\mathbf{v} \cdot \nabla)\mathbf{A},$$

and the force equation reduces to

$$\mathbf{F} = q\left[-\nabla(\phi - \mathbf{v} \cdot \mathbf{A}) - \frac{d\mathbf{A}}{dt}\right].$$

Now, consider the derivative of $(\phi - \mathbf{v} \cdot \mathbf{A})$ with respect to the velocity $\mathbf{v}$. Since the scalar potential is independent of velocity, we have

$$\frac{\partial}{\partial \mathbf{v}}(\phi - \mathbf{v} \cdot \mathbf{A}) = -\frac{\partial}{\partial \mathbf{v}}(\mathbf{v} \cdot \mathbf{A}) = -\mathbf{A}$$

and we have the last piece of the puzzle, namely,

$$-\frac{d\mathbf{A}}{dt} = \frac{d}{dt}\left(\frac{\partial}{\partial \mathbf{v}}(\phi - \mathbf{v} \cdot \mathbf{A})\right).$$

The Lorentz force is derivable thus from the velocity-dependent potential of the form $(\phi - \mathbf{v} \cdot \mathbf{A})$ by the Lagrangian recipe

$$\mathbf{F} = q \left[-\nabla(\phi - \mathbf{v} \cdot \mathbf{A}) + \frac{d}{dt}\frac{\partial}{\partial \mathbf{v}}(\phi - \mathbf{v} \cdot \mathbf{A}) \right]$$

and this leads to the all-important expression for the Lagrangian for charged particles in an electromagnetic field

$$L = T - q\phi + q\,\mathbf{A} \cdot \mathbf{v}.$$

Appendix 4: Fourier Decomposition of Field

The Klein–Gordon equation allows plane-wave solutions for the field $\phi(x)$ and it can be written as

$$\phi(x) = \frac{1}{(2\pi)^{3/2}} \int b(k) e^{ikx} dk$$

where $kx = k^0 x^0 - \mathbf{kr}$, $dk = dk^0 d\mathbf{k}$ and $b(k)$ is the Fourier transform that specifies particular weight distribution of plane-waves with different k's. Substituting the plane-wave solution into the Klein–Gordon, we get

$$\int b(k)(-k^2 + m^2) e^{ikx} dk = 0$$

indicating $b(k)$ to be of the form

$$b(k) = \delta(k^2 - m^2) c(k)$$

in which $c(k)$ is arbitrary. The delta function simply states that as the solution of Klein–Gordon equation, the plane-wave solution must obey the Einstein's energy-momentum relation, $k^2 - m^2 = 0$. The integral over dk therefore is not all over the $k^0 - \mathbf{k}$ four-dimensional space, but rather only over $d\mathbf{k}$ with k^0 restricted by the relation [for notational convenience we switch from $(k^0)^2$ to k_0^2]

$$k_0^2 - \mathbf{k}^2 - m^2 = 0.$$

Introducing a new notation

$$\omega_k \equiv +\sqrt{\mathbf{k}^2 + m^2} \quad \text{with only the } + \text{ sign,}$$

we have $k_0^2 = \omega_k^2$ and either $k_0 = +\omega_k$ or $k_0 = -\omega_k$. Integrating out k_0, the plane-wave solutions decompose into "positive frequency" and "negative frequency" parts. Using the identity

$$\delta(k^2 - m^2) = \frac{1}{2\omega_k}[\delta(k_0 - \omega_k) + \delta(k_0 + \omega_k)],$$

the plane-wave solutions become

$$\phi(x) = \frac{1}{(2\pi)^{3/2}} \int \frac{d^3\mathbf{k}}{2\omega_k} (c^{(-)}(-\omega_k, \mathbf{k})e^{-i\omega_k x_0}e^{-i\mathbf{k}\mathbf{x}}$$
$$+ c^{(+)}(\omega_k, \mathbf{k})e^{i\omega_k x_0}e^{-i\mathbf{k}\mathbf{x}}).$$

Changing $\mathbf{k}$ to $-\mathbf{k}$ in the first term, we have the decomposition

$$\phi(x) = \int d^3\mathbf{k}(a(\mathbf{k})f_k(x) + a^*(\mathbf{k})f_k^*(x))$$

where

$$f_k(x) = \frac{1}{\sqrt{(2\pi)^3 2\omega_k}}e^{-ikx} \quad \text{and} \quad f_k^*(x) = \frac{1}{\sqrt{(2\pi)^3 2\omega_k}}e^{+ikx}$$

After the decomposition into "positive frequency" and "negative frequency" parts, the notation k_0, as in e^{-ikx}, stands as a shorthand for $+\omega_k$, that is, after k_0 is integrated out, notation $k_0 = +\omega_k$.

Appendix 5: Mass Units for Particles

The units for mass that we are familiar with are grams, kilograms, pounds and even tons. Those units suffice to describe things that we deal with in our everyday life. Compared to the human-sized matter, the masses of the subatomic particles are so unimaginably miniscule that expressing them in terms of human-sized units is just not all that useful.

In terms of kilogram, the mass of an electron checks in at 9×10^{-31} kilograms. It is not unlike trying to express the size of a golf ball in terms of light-years. We need a new unit system more commensurable with the scale of the subatomic particles and it is not difficult to find one such unit system and it is based on the Einstein's mass-energy formula, $E = mc^2$.

The formula gives us way to express mass in terms of a suitable unit for energy divided by the square of the speed of light denoted by c. A suitable unit for energy for the world of subatomic particles is readily found in the unit called the electron-volt. One electron-volt of energy is the energy gained by an electron as it is accelerated across 1 volt in an electric field and corresponds to 1.602×10^{-19} joules, again meaninglessly miniscule when expressed in terms of the human-sized unit of energy. The symbol for the electron-volt is eV and thus mass can be expressed in terms of eV/c^2.

It turns out that a mass of $1\,\mathrm{eV}/c^2$ is way too small even by the standards of subatomic physics. In other words, we overdid it. We went too far down the scale, and more often than not we find it necessary to have to climb back up a little. So we use various multiples of electron-volt, such as keVs, kilo-electron-volts, MeVs, mega-electron-volts, GeVs, giga-electron-volts, TeVs, tera-electron-volts, and so on.

In these units, the mass of an electron becomes more "respectable." It checks in at $0.511\,\mathrm{MeV}/c^2$ while the mass of a proton is about $937\,\mathrm{MeV}/c^2$, often rounded up to $1\,\mathrm{GeV}/c^2$.

In the Natural Unit System (Appendix 1) where the speed of light c is set equal to 1, both energy and mass are expressed simply as eV, MeV, GeV and so on. It may possibly be a bit confusing, but they are completely interchangeable in the Natural Unit System.

Appendix 6: Mass-Range Relation

The three forces we are dealing with in the world of elementary particles — the electromagnetic, weak and strong — have vastly different ranges for the reach of their forces.

The electromagnetic force has an infinitely long range. Its force law is the well-known inverse square law (r^{-2}) and what is meant by an infinitely long range is that while the strength of the force falls off rapidly as the distance is increased, there is no specific point beyond which the force drops to zero. The tail of r^{-2} goes on forever, however weak it gets.

The weak and the strong forces operate entirely within the dimension of atomic nuclei and in fact have effective ranges that are fractions of the size of the smallest nuclei, those of hydrogen atom, to wit, a proton. The size of a proton is pegged at 10^{-15} meters and that is the range of the strong force. The range of the weak force is even shorter, about 10^{-18} meters. The strength of these forces simply drops to zero beyond their respective ranges.

There is an inverse relation between the range of a force and the mass of the particles that carry those forces. Photon is the carrier of the electromagnetic force and the weak bosons are the carriers of the weak force. Since the range of the weak force is so short, it tells you that the masses of the weak bosons must be large.

The simplest way to understand this mass-range inverse relation is by the energy-time uncertainty relation

$$\Delta E \Delta t \approx \hbar.$$

For ΔE we can use Mc^2 where M is the mass of the force carrying particle, photon and the weak bosons, and for Δt we can use R/c where R is the range of a force. We can immediately see the relation,

$$R \approx \frac{\hbar}{Mc}.$$

The range of the gravitation force is also infinite owing to the fact it obeys the inverse square law, just like the electromagnetic force. This is taken to imply that the gravitational force carrying particles, the gravitons, must also be massless. This statement, however, should be taken with some grains of salt. No one has been able to merge general relativity of Einstein with the quantum physics and the quantum field theory for the gravitational field does not yet exist. The graviton would be the quantum of the gravitational field if quantum gravity is successfully established.

Index

Printed in the United States
By Bookmasters